SOIL
MECHANICS AND
ENGINEERING

RONALD F. SCOTT
California Institute of Technology

JACK J. SCHOUSTRA
Converse, Davis and Associates, Pasadena

McGRAW-HILL BOOK COMPANY

New York   St. Louis   San Francisco   London   Toronto   Sydney

SOIL: Mechanics and Engineering

Library of Congress Catalog Card Number: 68-17195

ISBN 07-055798-5

2 3 4 5 6 7 8 9 10   (MAMM)   7 4 3 2

Preface

This book is intended to be a first course in the mechanics and engineering of soils. As such, the topics have been presented with a minimum of mathematical development. In the first six chapters, emphasis has been placed on the fundamental physics and mechanics of the subject so that the principles underlying the formulation of solutions to various classes of engineering problems can be clearly established in the last five chapters.

The theoretical and practical sections of the book were developed simultaneously so that they could be properly interrelated. It was the authors' intention, in doing this, that each practical area would refer to an appropriate section of theory, and that no topic would be enlarged on in the first section of the book without discussion in the second. To this end, the theoretical and practical sections were written by the senior and junior authors respectively, and then exchanged for extensive revision and rewriting.

Because of the increasing use of computers in soil engineering the material has been presented, in general, in a form most suitable for numerical calculations. This enables certain simplified solutions to be developed by hand calculations, so that understanding of the physical processes is enhanced. The analogous techniques for solution of more complex problems by computer follow logically.

The problems in the practical section reflect typical situations confronting practicing soil engineers. The entire text is aimed at demonstrating what often has been left in doubt: that theoretical and practical soil mechanics are well matched to solve everyday soil engineering problems.

Ronald F. Scott
Jack J. Schoustra

Contents

PART 1 | PHYSICS AND MECHANICS OF SOILS

Soils as they occur in nature are heterogeneous accumulations of solid grains, water, and gas. According to the sizes and composition of the grains and the relative proportions of the various components, soils exhibit a very wide range of properties. At any one engineering site the ground to be loaded by a structure will inevitably consist of a number of layers or regions of different soil materials, all exhibiting different properties. On occasion the variations will be extreme. It is the task of the soil engineer to calculate the displacements, settlements, stresses, and forces which will be caused in the soil mass by the proposed structure and to make sure that none of these exceed certain limits. If they do, the structure may be distorted, exhibiting unsightly cracks, or become functionally unusable because of the severance of connecting utilities, or in extreme cases fail or collapse.

Up to the twentieth century, the behavior of soils was largely unknown and soil investigational methods both in the field and in the laboratory were very crude. In the first decades of the twentieth century the physics and mechanics of soil behavior were studied and clarified, so that some understanding of their complexity was developed. Thus, in certain simple cases where a soil profile was relatively homogeneous, reference could be made to idealized mathematical models for calculations of displacements and stresses. However, mathematical difficulties prevented the analysis of more complex models—more complex either in their geometry or in the nature of the material behavior treated—and the practicing engineer was compelled to rely largely on his judgment in relating a given practical situation to a simplified model.

The development and availability of large digital computers have vastly enlarged the scope of the soil problems for which solutions can be obtained. It is now possible to model the *geometry* of a soil engineering situation as well as it can be determined from the field investigation. However, knowledge of the soil properties with which the various layers or regions are endowed has still to be extended before field problems can be successfully solved in all cases.

It is the purpose of the first six chapters to establish the basic physics and mechanics of soil behavior and to illustrate the developments with idealized analytical models where appropriate. However, emphasis is placed primarily on numerical approaches and formulations rather than on mathematical techniques, since it is the former which lead directly into the employment of computing machines for solutions on an individual case basis. In the last five chapters the application of these fundamentals to typical soil engineering problems will be treated.

1 | **Nature of Soils**

To the soil engineer, soils are the unaggregated or uncemented deposits of mineral and/or organic particles or fragments covering large portions of the earth's crust. The range in particle sizes in a soil or from one soil to another is enormous; it may extend from grains only a fraction of a micron (10^{-4} cm) in diameter up to boulders many feet in diameter. As we shall see, soils are primarily classified on the basis of their grain size and their distribution of grain sizes. Any one soil may be composed of mineral particles all of approximately the same size; on the other hand, it may consist of particles of a very wide range of sizes.

The individual particles themselves may consist of one mineral or mineral aggregates, and they may take any shape between the extremes of plate- or needle-shaped particles and particles of roughly spherical form. The term "soil" as we use it here includes the mineral particles forming a structural skeleton or framework of the solid mass and their pores, which may be filled with water or other liquids and gas bubbles. "Dry" soil will contain only small amounts of water. Most soils with which we deal are comprised of predominantly mineral particles, but some few are derived from organic materials such as peats or mosses and, on occasion as in refuse dumps, other organic matter.

Relatively unusual soils which may be encountered in foundation engineering may also include iron ore, coal, sugar, flour, wheat or other grains, or salt or other powdered chemicals. All such materials have many similar properties which may be analyzed by the methods of soil mechanics as explained in this book. We consider soils to be uncemented in order to exclude from our present considerations materials such as asphalt, concrete, sandstone, or other relatively strongly cemented materials, but in practice, especially in arid areas, it is common to find soil deposits

3

which are relatively weakly cemented by salt deposits at the points of contacts between grains. The bonds formed by the salts in these materials are relatively easily disrupted, and such materials are considered as soils in our understanding. As a result of the weight of immensely thick ice sheets in periods of glaciation, certain soils termed *glacial tills* have been subjected to high pressures and are found in nature in extremely dense and compact states, but they are still referred to in the present context as soils. We shall proceed to examine soils in more detail in the next section.

1-1 ORIGIN, IDENTIFICATION, AND CLASSIFICATION

Soils, with the exception of the soil-like materials listed above, are derived from the breakdown of the massive rocks constituting the earth's crust. The rocks themselves may be the original material which has crystallized out of the mass of molten magma ejected from the interior of the earth into or onto the crust or which may have developed from the subsequent alteration of the original rocks as a result of pressure and temperature changes. The basic rocks themselves may be broken down into soils which, in turn, may be subjected to high pressures and temperatures to form rocklike materials again in the form of *sedimentary rocks*. If the temperatures are sufficiently high, the individual particles may lose their identity in a remelted mass to form a *metamorphic rock*. In the processes forming soils and rocks in the earth's crust, rocks and soils may be formed alternately many times and periodically broken down and reaggregated.

Many processes are involved in breaking down a rock to form soil. Whether it is formed deep below the earth's crust with a large overburden of other rocks or sediments or on the surface as a result of an outpouring of molten lava or magma, a rock mass as it cools develops stresses within it. These stresses are a result of uneven cooling and also a result of the distortions and deformations the mass undergoes through the movement of adjacent rocks or surfaces by tectonic action. Consequently, to some extent, all rocks are fractured or jointed as a result of the stresses exceeding the strength of the rock in various local positions. The pattern of fracturing and jointing frequently reflects both the stress distribution and the crystalline formation of the rock materials.

On eventual exposure at the surface of the earth, a mass of jointed and fissured rock will be subjected to the physical action of water, wind, and gravity, so that blocks or sections will become detached from the parent body and move to new equilibrium positions of their own. The movement will generally be accompanied by further fracturing or fissuring of each individual fragment when it comes in contact with other rocks or fragments as it is moved or falls as a result of wind, water, or frost action in the original fissures. Thus, any rock mass tends to gradually break down into smaller fragments and form soils as it does so. This process is called *weathering,* and it results in finer and finer soil fragments. The more these fragments come in contact with one another as a result of rolling down slopes, being moved along stream beds or rivers, or being blown along desert surfaces by the winds, the more rounded in general the fragments become.

Usually the natural processes of wind or water action tend to sort the particles in such a way that particles of one range of sizes will tend to congregate in a particular location and the finer or coarser particles will be deposited or moved elsewhere. Thus, fairly frequently, relatively uniform wind-blown or water-laid deposits of soils are encountered.

In the particular physical process of glacier action, a moving glacier will pick or pluck up rocks underlying the ice sheet and carry them along in the ice mass, grinding them against the parent or other rocks as motion continues. Such action will result in soil of a wide range of grain sizes from very fine material, sometimes called rock flour, up to the very largest boulders. These materials can be transported many miles by glacier action, and frequently they are ultimately deposited in a melt-water lake or river at the melting edge of the glacier. Such deposits in the form of *moraines*, *eskers*, and *drumlins* may in turn be overrun by glaciers in subsequent advances of the ice sheet. They are also, after deposition, eroded by wind or water to form new deposits in lakes or deltas or in dunes elsewhere.

Soil formation, by whatever process, is continuous, and we, as soil engineers, see what amounts to a single frame of an extremely long movie of soil and rock dynamics. The grinding or comminuting action of one soil grain upon another results from the forces and stresses generated in the grains as they are flung into contact with one another upon falling or being tumbled about in air or water. The force at contact therefore depends upon the mass and acceleration of the grain due to the hydraulic stresses acting upon it.

As the grains get smaller and smaller, the mass of the individual grains diminishes to such an extent that in ordinary physical processes, for minerals of a given strength, a limit is reached at which the impact stresses are not sufficient to cause fracture in individual grains. Thus, the physical process of soil formation has a lower limit to grain size (with the particular exception of the fine grains produced by the grinding action of glaciers). If no other processes were at work, then on the average the finest grains in the soils produced in nature would have a diameter of some tens of microns. However, other processes *are* at work.

If we consider a mineral grain of cubic shape and side length 1 cm, we can calculate that it has a volume of 1 cm³ and a surface area of 6 cm². If this grain is flung or thrown against a grain of similar size, one or other of the grains may be broken and the area of new surface formed by the break will be approximately proportional to the original surface area of the grains, whereas, as we have seen above, the forces or stresses at impact are proportional to the mass or volume of the grains. For grains of any one kind, the lower the ratio of the surface area to the volume of the grains, the more fractures will be produced, since this implies a higher ratio of forces or stresses to new surface developed. For a cubic particle 0.1 cm on a side the volume is equal to 10^{-3} cm³, whereas the surface area equals 6×10^{-2} cm². Thus for the 1-cm particle, the ratio of surface area to volume equals 6 cm⁻¹, whereas for the 1-mm particle the ratio of surface area to volume is 60 cm⁻¹. That is to say, the ratio of surface area to volume (hereafter called the *specific surface*) varies inversely as the linear size of the particle.

Now, if we were to place soil grains in a liquid whose chemical properties were such as to dissolve the minerals of which the soil is composed, the chemical would be able to attack or combine with only the molecules at the surface of each particle and would dissolve an amount of the mineral in proportion to the ratio of the surface area to the volume, or, in other terms, in proportion to the number of molecules occupying the surface of a particle as compared to the number of molecules situated internally in the particle. Consequently, the effectiveness of chemical action *increases* as the specific surface of the individual particles goes up, and consequently as the size of the particles goes down. Proportionately, therefore, the solution action of rain water falling upon massive rocks is unimportant as a particle-size reducing agency in comparison with the processes of solution at work as water flows through the pores of a relatively fine-grained soil.

Chemical reactions between the acids dissolved in water and the soil solids give rise to the solution of minerals in the soil grains; these minerals in solution will recombine and recrystallize under differing conditions of temperature or pressure at other points in the water-flow process to give rise to new minerals. These new minerals consist of different patterns or arrangements of the molecules in the original minerals.

The new patterns tend to give rise to small mineral particles which are needle- or plate-shaped and have lengths or diameters tens to hundreds of times their thickness in contrast to the predominantly equidimensional or bulky proportions of the grains formed of the original rock minerals by physical processes. The new minerals formed by chemical action are known as the *clay minerals*, and the particles have sizes from hundredths of a micron to tens of a micron in diameter.

When carried by moving water to a lake or the ocean, the small size of the particles causes them to take a long time to settle out of the water to form deposits of fine-grained materials on the beds of lakes and other bodies of water. In the cases of lakes or the sea fed by rivers bringing down a wide range of soil deposits, the clay minerals may be deposited at the same time as a large variety of other sizes of mineral particles to give rise to soil deposits of wide size distributions.

The clay beds themselves, as is the case with other soil deposits, may subsequently be uplifted and reeroded by the forces of nature to form successive soil deposits of various characteristics depending on the material's history of chemical action, deposition, wetting, drying, and stress variation. With the wide range of processes at nature's disposal it will be recognized that the variety of soils produced will have a spectrum of deformational and other properties to match their varieties.[1*]

In soil mechanics and soil engineering we are chiefly interested in the behavior of soils when they are subjected to applied stresses and forces as a result of the erection of structures on or in the ground and the excavation of cuts, tunnels, or shafts in the ground. Important factors in all soil behavior are the quantity of water present in the pores of the soil and the amounts of water which flow through the pores under various conditions. The range of soil behaviors with respect to these considerations will be discussed in future chapters. At this stage it is useful to distinguish between the various soils on a preliminary basis for suitable classification as to their mechanical behaviors later, and we therefore proceed to consider the identification and classification of soils.

The physical and chemical processes which result in a gradation of sizes of granular materials from immense boulders down to tiny particles composed of clay minerals also suggest that this will be a convenient method of classifying the material, and most systems of classification which have been devised are, in fact, based upon grain sizes. We are interested in both the approximate mean size of a given soil and also the range of sizes which is present, and we find in practice that important mechanical property distinctions and differences develop at a soil grain size approximately at which the chemical and physical processes are also separated. In other words, certain material behaviors may be associated with the coarser grain sizes and other qualitatively different behaviors with soils composed of the finer grain sizes.

Consequently, it is of interest to determine in any given soil sample the proportions of relatively coarse and relatively fine material present, and we may do this by arbitrarily considering that soil coarser in size than the mesh openings of a No. 200 sieve will be termed *coarse* in our classification system and smaller-grained soil will be termed *fine*. The openings in a No. 200 sieve are 74 microns square; we shall later see what type of soil this size corresponds to. For the coarser-grained material, sieves serve a useful purpose for determining the proportions of the soil consisting of grains of various sizes.

The classification of soil by grain-size distribution is accomplished by setting up a nest or stack of sieves in which each sieve is set above a second one whose mesh opening is commonly half the size of the opening in the first. Frequently, seven or eight such sieves are used; they range in size from perhaps a $\frac{3}{4}$-in. opening down to the No. 200 sieve with the 74-micron opening. However, the selection of the sieves is usually based on the observed size range in the soil in order to describe the variation of sizes present most accurately. A known weight of soil is added to the top of the nest of sieves; the nest is shaken vigorously for 10 or 15 min; and the weight of the soil retained on each sieve is measured. The soil on any one

* Superscript numbers are those of the references listed at end of chapter.

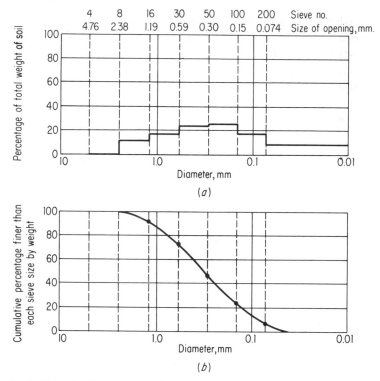

FIG. 1–1   *Distribution of particle sizes in a soil. (a) Particle size distribution. (b) Cumulative particle size distribution.*

sieve is then of such size as to pass through the openings in the sieve resting upon it. A pan catches the grains passing the No. 200 sieve.

With the information provided by a sieve analysis, the distribution by weight of grain sizes in the soil under study can be plotted on a graph such as shown in Fig. 1–1. It will be noted that the grain-size scale in Fig. 1–1 is logarithmic because of the extremely wide variation of grain sizes present in most soils, frequently covering a range of several to many orders of magnitude. Two soils can therefore best be compared by referring them to a logarithmic scale of grain size as shown. If the percentages of the soil by weight as shown plotted in Fig. 1–1a are summed or accumulated from the left or from the right, whichever is preferable, the cumulative curve of grain-size distribution shown in Fig. 1–1b is obtained. As Fig. 1–1b is drawn, it is apparent that the cumulative percentage by weight refers to the percentage of the soil finer than each sieve at which a point is plotted. Summing the data the other way would give a curve of cumulative percentage coarser than each given sieve size.

When a test like this is carried out, it becomes apparent that it is difficult to get the soil to pass through the finer sizes of sieve opening partly because of aggregation of the finer material in dried lumps, so that it is usual to wash the material through by water before drying the quantities on each sieve. It will be understood, therefore, both from an examination of the construction of the finer sieves and the difficulty of sieving the material, that sieve analysis would not be practicable for soils finer than those accumulating on the No. 200 sieve. In practice, other methods of grain-size analysis are available, and they will be described below.

We have described the soil shown in Fig. 1–1 by reference to the sizes of various sieves without assigning names to the different ranges of soil sizes. It is convenient at this stage to consider how we may describe soils and how, in consequence, they may be classified on the basis of size. Many different size classifications of soils have been proposed, but the one that is most commonly used among soil engineers at present is referred to as the MIT grain-size classification, and it is widely employed because of the ease of memorizing it.

The MIT system is shown in Table 1–1, in which it is seen that the different boundaries between grain sizes occur at intervals of the digits 2 and 6. Consequently, it is only necessary to remember that this is so and to remember the separating size at any one interval in Table 1–1. It is convenient, for example, to remember that the upper limit of coarse sand is 2 mm or, conversely, that this represents the lower limit of gravel size. With this in mind, the table can be constructed by writing 2, 0.6, 0.2, etc., alternately in diminishing size steps. It will be noted that the assignment of names to the various sizes corresponds to our subjective impressions of material and to the names that are used with less precision in everyday life. That is to say, one would not normally term a material "sand" were the sizes of the individual grains half an inch in diameter.

On looking at the table one can also observe that the lower limit of the sand size occurs at 0.06 mm, which is very close to the size of the opening of a No. 200 sieve, and that, as a consequence, the sieve used in the manner described above separates sands and gravels, the coarser-grained soils, from silts and clays, the finer-grained materials. We do not usually distinguish among the different grades of coarse, medium, and fine clays. Table 1–1 has been employed to construct the scale with names of materials in Fig. 1–2.

TABLE 1–1 MIT* GRAIN–SIZE CLASSIFICATION SYSTEM

Grain Size, mm							
2	0.6	0.2	0.06	0.02	0.006	0.002	
	Coarse	Medium	Fine	Coarse	Medium	Fine	
Gravel	Sand			Silt			Clay

* Massachusetts Institute of Technology.

It is usual in this regard to give the soil the general name after that size of material which constitutes the larger percentage of the soil sample studied, and we see, for example, in Fig. 1–1 that we would describe this material as a sand. The size constituting the second largest fraction of material present is usually employed as a descriptive adjective qualifying the primary name of the material. Thus the material of Fig. 1–1 might be classified as a silty sand. The other names of soils, such as gravelly sand and clayey silt, follow logically, and with a little practice one can visualize their location on graphs such as Fig. 1–1. When the soil contains a wide range of grain sizes, again as shown in Fig. 1–1, it may be referred to as a well-graded soil; if one size of material predominates, the soil is referred to as uniform.

Various soil types are shown graphically in Fig. 1–2, along with the accompanying adjectives (based on the MIT size system) which would be used in engineering or boring reports. A precise definition of a soil must always be based on graphs such as Fig. 1–1 or 1–2. In practice, however, constant handling and examining of soil with reference to such graphs generally provide sufficient experience for identifying the type by visual inspection. If the nature of the material is important for any reason, a grain-size analysis should always be performed to reinforce the subjective description.

Since fine-grained soils, such as silts and clays, cannot be analyzed by the sieve method, other techniques must be adopted. The most common one in use at present is the so-called hydrometer method. In this technique a small weight of the fine-grained soil, which is usually found in a moist condition, is mixed with water in a large glass graduate, and a chemical called a dispersant is usually added to ensure that the individual grains of soil separate from one another and can therefore be correctly measured. If the mixture or suspension of soil and water is shaken vigorously and the glass graduate is then placed on its base on a bench, the various sizes of soil particles will proceed to settle at rates according to their size. As a consequence, the initially uniform density of the suspension will begin to vary from top to bottom, becoming more dense at the bottom of the graduate where the grains are accumulating, and less dense at the top. At any given level in the suspension the density will vary with

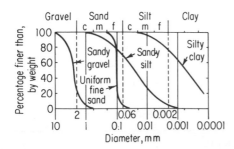

FIG. 1–2   *Various soil-size distribution curves.*

time in a way dependent upon the size distribution of the soil grains present. Therefore, a hydrometer that is inserted into the suspension at intervals in effect measures the specific gravity or density of the fluid at a given depth.

A mathematical relationship known as Stokes' law tells us the relationship among the size, specific gravity of the particles and of the fluid, and the velocity of a particle. We can apply this to the readings of the hydrometer with time to enable the percentage of the particles by weight smaller than a certain given size to be calculated for each time the hydrometer is lowered into the suspension. This enables us, therefore, to calculate grain-size distribution curves of the type of Figs. 1–1 and 1–2 for these fine-grained materials. Very small soil grains take a long time to settle out of suspension, and the hydrometer method therefore becomes somewhat inconvenient for distinguishing among grain sizes smaller than about 0.002 mm (2 microns). To discriminate among grains smaller than this size would require the use of centrifuge methods, or alternatively, the individual grains could be inspected by employing an electron microscope. Although this method does not give a picture of the distribution of grain sizes present, such a distinction becomes relatively unimportant in the finer materials, and frequently the shape and proportions of individual grains are of more interest than their size.

With the finer-grained materials most engineering interest centers in their mechanical behavior at different water contents,* and as a consequence it becomes more convenient to measure the qualitative mechanical behavior of the finer-grained materials by means of simple empirical mechanical tests. The tests which are employed for this purpose were devised by a Swedish chemist connected with the clay and pottery industries in Sweden at the beginning of the twentieth century, and they are called the Atterberg limit tests after him. He observed that over a range of water contents a clay exhibits a characteristic variation of behaviors from that of essentially a liquid at extremely high water contents to that of a brittle solid when most of the water is absent. He devised tests, among others, to distinguish when the clay behavior changed from that of a viscous liquid to that of a plastic or moldable solid and also to distinguish when the material behavior changed from plastic to brittle as the clay dried out. His tests are therefore called the *liquid limit test* (referring to the lower limit of liquid behavior), and the *plastic limit test* (referring to the lower limit of plastic behavior).

A soil is considered to behave in a plastic way when it can be molded or worked and will retain a new shape impressed upon it without either returning to its original shape or fracturing. This is a characteristic of many clays in a certain range of water content, and it is, of course, the behavior we normally attribute to modeling clay and clay used in the

* The water content of a soil and other descriptive parameters are defined later in this chapter.

manufacture of pottery. If one squeezes or distorts the material in one's hand, it will retain the distorted shape. Atterberg tests were employed by Terzaghi and subsequently modified by Casagrande for the purpose of describing the appropriate engineering behavior of a given clay. In the Atterberg limit test the clay is reworked with water and placed in the cup of a simple device called the Atterberg limit device and a groove is cut in the clay. The cup is then rapped on a hard surface in such a way as to tend to close the sides of the groove. This rapping imposes small shearing stresses on the clay. An arbitrary value of 25 blows is taken to represent a state of the clay transitional between liquid and plastic behavior. If the gap in the material does not close below 25 raps or blows of the cup, the material is considered to be still plastic. If after more water is added to the clay the gap closes at fewer than 25 blows, the material is considered to have become liquid.

The water content is the important measurement in the Atterberg test, and we therefore assign the name *liquid limit water content* to the water content of the material at which the groove or gap just closes at 25 blows. It will be realized, of course, that the number 25 is arbitrary, and the method of rapping the cup and manufacturing the groove or gap in the clay soil are selective variables; were any one of them changed, the value of the liquid limit water content would be changed. The test is therefore an empirical one, and the results depend subjectively on the way the technician carries out the test, the preparation of the soil, and many other variables connected with the actual operation of the equipment. However, if it is carried out in a consistent way, the test gives a valuable indication of the type of soil tested and its mechanical properties.

The *plastic limit test* is even simpler in operation and requires no equipment other than the means of determining the water content of the soil. It is intended to determine the water content at which the clay just fails to behave plastically but cracks in a brittle fashion under distortional stresses. The method is to take small rolls of the clay, $\frac{1}{8}$ in. in diameter, at progressively lowering water contents and remold them between the hand and a glass plate until a water content is reached at which the soil becomes difficult to roll; instead of forming a smooth cylinder, the soil cracks and splinters. With a little practice it becomes quite simple to determine this stage of behavior consistently. The water content at which cracking of the soil roll occurs is called the *plastic limit water content*. The range of water contents over which the clay behaves plastically is also of interest, and it is represented by the difference between the liquid limit and plastic limit water contents. This value, as a percentage, is called the *plasticity index* of the soil.

If one were to take a cube or cylinder of clay and carefully measure its volume as the water content was steadily reduced from a value in the region of the liquid limit, one would find a volumetric water content behavior as shown in Fig. 1–3. As the water content was reduced in the

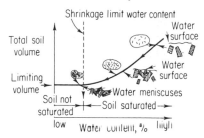

FIG. 1–3   *Volume change on drying.*

early stages, the overall volume of the soil would decrease until eventually a lower limiting volume (at which the water content was not zero) was reached and no further shrinkage in volume would take place. This behavior is usually explained by considering that, at the very high water content, the soil particles are separated from one another, the spaces in between being filled with water. Consequently, as water is removed, the soil particles move closer together until at some stage they form a closely intermeshing network of particles which are able to take the stresses imposed by the meniscuses developing at the surfaces. The stages of the behavior are shown in Fig. 1–3 by the diagrammatic sketches accompanying the graph. (See Chapter 2 for an explanation of stresses due to meniscuses.) When the close spacing between particles has been achieved, the volume change virtually ceases. The water content which corresponds to the volume of soil at which no further change in volume occurs is called the *shrinkage limit water content;* it is rarely employed in practice except in situations where the drying out or shrinkage of clay soils may influence the foundations of a structure.

We now have a major portion of the information necessary to classify any soil which may be encountered in practice. On receiving the soil from an excavation or boring, the sample may be separated into the coarser and finer fractions on a No. 200 sieve, and a sieve analysis may be carried out on the coarser sizes for the purpose of plotting diagrams such as Figs. 1–1 and 1–2 for the material. The finer fraction of soil may be subjected to a hydrometer analysis, and this can either be added to the information from the sieve analysis on the coarser fraction or plotted separately on another grain-size distribution chart to give the grain-size picture of the soil. The finer fraction will also generally be subjected to the liquid and plastic Atterberg limit tests, since the compressibility, permeability, and strength of the soils are generally indicated by those tests.

Many different classification systems have been devised to take into account all of the information in the above analyses for the purpose of describing the engineering behavior of the soil qualitatively. However, the principal one in use today was established during World War II, and it is now termed the unified classification system (UCS). It is set out in detail in Table 1–2. The notes accompanying the table explain the simple

(*Continued on page 18 at end of table note.*)

TABLE 1-2 UNIFIED CLASSIFICATION SYSTEM

PART 1 COARSE-GRAINED SOILS

(More than half of material is larger than No. 200 sieve size.[a])

Soil	Major Divisions	Group Symbols[b]	Typical Names	Field Identification Procedures[c]
Gravels[d]	Clean gravels[e]	GW	Well-graded gravels, gravel-sand mixtures, little or no fines	Wide range in grain sizes and substantial amounts of all intermediate particle sizes
		GP	Poorly graded gravels or gravel-sand mixtures, little or no fines	Predominantly one size or a range of sizes with some intermediate sizes missing
	Gravels with fines[f]	GM	Silty gravels, gravel-sand-silt mixture	Nonplastic fines or fines with low plasticity (for identification procedures see ML in part 2 of table)
		GC	Clayey gravels, gravel-sand-clay mixtures	Plastic fines (for identification procedures see CL in part 2 of table)
Sands[g]	Clean sands[e]	SW	Well-graded sands, gravelly sands, little or no fines	Wide range in grain size and substantial amounts of all intermediate particle sizes
		SP	Poorly graded sands or gravelly sands, little or no fines	Predominantly one size or a range of sizes with some intermediate sizes missing
	Sands with fines[f]	SM	Silty sands, sand-silt mixtures	Nonplastic fines or fines with low plasticity (for identification procedures see ML in part 2 of table)
		SC	Clayey sands, sand-clay mixtures	Plastic fines (for identification procedures see CL in part 2 of table)

[a] All sieve sizes in this table are U.S. standard.

[b] Soils possessing characteristics of two groups are designated by a combination of group symbols; for example, GW-GC is well-graded gravel-sand mixture with clay binder.

[c] Excluding particles larger than 3 in. and basing fractions on estimated weights.

[d] More than half of coarse fraction is larger than No. 4 sieve size.

[e] Little or no fines.

[f] Appreciable amount of fines.

[g] More than half of coarse fraction is smaller than No. 4 sieve size.

Source: Table adapted from U.S. Army Corps of Engineers Waterways Experiment Station, The Unified Soil Classification System, *Tech. Memo* 3-357, March, 1953.

Laboratory Classification Criteria
(*see Plasticity Chart in part 2 of the table*)

$C_u = \dfrac{D_{60}}{D_{10}}$ greater than 4 where C_u = uniformity coefficient; D_{60} (etc.) = grain

size than which 60% of the soil is finer

GP does not meet all gradation requirements for GW

Atterberg limits below *A* line or PI less than 4	Above *A* line with PI between 4 and 7 are *borderline* cases requiring use of dual symbols
Atterberg limits above *A* line with PI greater than 7	

$C_u = \dfrac{D_{60}}{D_{10}}$ greater than 6

SP does not meet all gradation requirements for SW

Atterberg limits below *A* line or PI less than 4	Limits plotting in hatched zone with PI between 4 and 7 are *borderline* cases requiring use of dual symbols
Atterberg limits above *A* line with PI greater than 7	

Information Required for Describing Soils

Give typical name; indicate approximate percentages of sand and gravel, maximum size; angularity, surface condition, and hardness of the coarse grains; local or geologic name and other pertinent descriptive information; and symbol in parentheses.

For undisturbed soils add information on stratification, degree of compactness, cementation, moisture conditions, and drainage characteristics.

Example. *Silty sand*, gravelly; about 20% hard, angular gravel particles $\frac{1}{2}$ in. maximum size; rounded and subangular sand grains, coarse to fine; about 15% non-plastic fines with low dry strength; well compacted and moist in place; alluvial sand; (SM).

Determine percentages of gravel and sand from grain-size curve. Depending on percentage of fines (fraction smaller than No. 200 sieve size) coarse-grained soils are classified as follows:

Less than 5%	GW, GP, SW, SP
More than 12%	GM, GC, SM, SC
5% to 12%	*Borderline* cases requiring use of dual symbols

TABLE 1-2   UNIFIED CLASSIFICATION SYSTEM

PART 2   FINE-GRAINED SOILS

(More than half of material is smaller than No. 200 sieve size.[a])

Soil	Major Divisions	Group Symbols[b]	Typical Names	Identification Procedures on Fraction Smaller than No. 40 Sieve Size[c]		
				Dry Strength	Dilatancy	Toughness
Silts and clays	Liquid limit less than 50	ML	Inorganic silts and very fine sands, rock flour, silty or clayey fine sands or clayey silts with slight plasticity	None to slight	Quick to slow	None
		CL	Inorganic clays of low to medium plasticity, gravelly clays, sandy clays, silty clays, lean clays	Medium to high	None to very slow	Medium
		OL	Organic silts and organic silty clays of low plasticity	Slight to medium	Slow	Slight
	Liquid limit more than 50	MH	Inorganic silts, micaceous or diatomaceous fine sandy or silty soils, elastic silts.	Slight to medium	Slow to none	Slight to medium
		CH	Inorganic clays of high plasticity, fat clays	High to very high	None	High
		OH	Organic clays of medium to high plasticity, organic silts	Medium to high	None to very slow	Slight to medium
Highly organic soils		Pt	Peat and other highly organic soils	Readily identified by color, odor, spongy feel and frequently by fibrous texture		

[a] All sieve sizes in this table are U.S. standard.

[b] Soils possessing characteristics of two groups are designated by a combination of group symbols; for example, GW-GC is well-graded gravel-sand mixture with clay binder.

[c] See detailed Field Identification Procedures for Fine-grained Soils or Fractions on facing page.

Information Required for Describing Soils

Give typical name; indicate degree and character of plasticity; amount and maximum size of coarse grains; color in wet condition; odor, if any; local or geologic name and other pertinent descriptive information; and symbol in parentheses.

For undisturbed soils add information on structure, stratification, consistency in undisturbed and remolded states, moisture and drainage conditions.

Example. *Clayey silt*, brown; slightly plastic; small percentage of fine sand; numerous vertical root holes; firm and dry in place; loess; (ML).

Plasticity Chart for Laboratory Classification of Fine-grained Soils

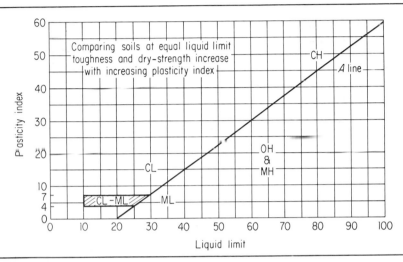

Field Identification Procedures for Fine-grained Soils or Fractions

These procedures are to be performed on the minus No. 40 sieve size particles, approximately $\frac{1}{64}$ in.

Dry strength (crushing characteristics). After removing particles larger than No. 40 sieve size, mold a pat of soil to the consistency of putty, adding water if necessary. Allow the pat to dry completely by oven, sun, or air-drying, and then test its strength by breaking and crumbling between the fingers. This strength is a measure of the character and quantity of the colloidal fraction contained in the soil. The dry strength increases with increasing plasticity.

High dry strength is characteristic for clays of the CH group. A typical inorganic silt possesses only very slight dry strength. Silty fine sands and silts have about the same slight dry strength, but can be distinguished by the feel when powdering the dried specimen. Fine sand feels gritty, whereas a typical silt has the smooth feel of flour.

Dilatancy (reaction to shaking). After removing particles larger than No. 40 sieve size, prepare a pat of moist soil with a volume of about one-half cubic inch. Add enough water if necessary to make the soil soft but not sticky.

Place the pat in the open palm of one hand and shake horizontally, striking vigorously against the other hand several times. A positive reaction consists of the appearance of water on the surface of the pat which changes to a livery consistency and becomes glossy. When the sample is squeezed between the fingers, the water and gloss disappear from the surface, the pat stiffens, and finally it cracks or crumbles. The rapidity of appearance of water during shaking and of its disappearance during squeezing assist in identifying the character of the fines in a soil.

Very fine clean sands give the quickest and most distinct reaction, whereas a plastic clay has no reaction. Inorganic silts, such as a typical rock flour, show a moderately quick reaction.

Toughness (consistency near plastic limit). After particles larger than the No. 40 sieve size are removed, a specimen of soil about one-half inch cube in size is molded to the consistency of putty. If too dry, water must be added, and if sticky, the specimen should be spread out in a thin layer and allowed to lose some moisture by evapo-

(*Continued on next page.*)

ration. Then the specimen is rolled out by hand on a smooth surface or between the palms into a thread about one-eighth inch in diameter. The thread is then folded and rerolled repeatedly. During this manipulation, the moisture content is gradually reduced and the specimen stiffens, finally loses its plasticity, and crumbles when the plastic limit is reached.

After the thread crumbles, the pieces should be lumped together and a slight kneading action continued until the lump crumbles.

The tougher the thread near the plastic limit and the stiffer the lump when it finally crumbles, the more potent is the colloidal clay fraction in the soil. Weakness of the thread at the plastic limit and quick loss of coherence of the lump below the plastic limit indicate either inorganic clay of low plasticity, or materials such as kaolin-type clays and organic clays which occur below the *A* line.

Highly organic clays have a very weak and spongy feel at the plastic limit.

---

tests which are necessary to determine the characteristics of a soil, in addition to the grain-size analysis, for the purpose of classifying it. It will be seen that in addition to the gravel, sand, silt, and clay sizes of Table 1–1, an additional category of *organic materials* has been added and that, with the above information, the majority of soils can be classified in an abbreviated form employing only a two-letter combination. Experience has taught us to relate the various written properties in the table of the UCS to the appropriate materials.

1-2 SOIL MINERALS

In the rocks of the earth's crust various combinations of the basic elements, oxygen, hydrogen, silicon, aluminum, potassium, calcium, iron, etc., in regular crystalline arrangements give rise to different minerals. The nature and proportion of the minerals present in a rock, in addition to the size and distribution of the mineral crystals present, are used to classify the rock. Thus, gravels, sands, and to some extent silts, all being formed by the abrasion and comminution of the rocks of the earth's crust, may consist of grains composed of one or a number of mineral crystals. Some soils in the coarser range are composed of surprisingly uniform deposits both as to size and the minerals present because of the mechanism of sorting by wind and water action. One such well-known soil is the so-called Ottawa (Illinois) sand, which is derived from a weakly cemented sandstone and consists almost entirely of quartz grains (which are formed from the elements silicon and oxygen). Calcium carbonate sands derived from the breaking up of shell fragments on beaches may also be found.[1]

When we compare the mechanical behaviors of sands and gravels composed of different typical rock minerals or mineral aggregates, we find that the behavior of the sand depends primarily on its state of denseness or looseness and on the size and shape of grains present, and not on the nature of the minerals present in the soil. This occurs, because, at the pressures usually involved in soil engineering applications, most soil minerals (having already been subjected to considerable amounts of impact and abrasion during the soil-forming processes) are relatively strong and do not fracture or disintegrate. The weaker or softer minerals are generally removed either mechanically or chemically during the soil-forming processes, and therefore sands composed of relatively soft mineral particles are comparatively rare. However, considerations of particle breakdown may be of importance in the analysis of materials that are of a granular nature but are not composed of soil minerals, such as those listed in the early part of this chapter. Consequently, in an analysis of granular soil as it might relate to mechanical behavior, the nature of the minerals or mineral aggregates present is usually unimportant. Thus a mineral analysis gives no information that is pertinent to the soil *mechanics* of coarse-grained soils, and it is not usually carried out.

The situation is otherwise in the fine-grained soils composed of clay minerals. The clay minerals are produced by the recrystallization and recombination of the original rock-forming minerals in solution, and the particles formed of these new minerals have characteristically different size ranges and behaviors, depending on the crystalline form attained.[2] A knowledge of the principal different minerals is of interest and value in assessing the probable behavior of a fine-grained soil, and frequently mineral analyses are carried out on materials whose behavior is to be interpreted.

There are three main groups of the clay minerals of interest to soil engineers: the kaolinites, the montmorillonites, and the illites; these will be discussed in turn. Two basic "building blocks," or crystalline units, go into the formation of a clay mineral, and combination of the two units in different ways gives rise to the three clay mineral groups. One building block is the mineral silica, or silicon dioxide, with the formula SiO_2, whose atoms are arranged in the form of a tetrahedron; aggregates of these tetrahedral molecules tend to form sheetlike or platelike minerals as in the micas. The other building block is a mineral called gibbsite, an aluminum hydroxide with the general formula $Al(OH)_2$. The atoms in a gibbsite molecule arrange themselves in an octahedron, and this also gives rise to sheetlike structures. The distance between the molecules is very similar to the spacing between certain molecules in the silica sheet. In Fig. 1–4a the two building blocks are shown schematically.

Because of the similarity of the spacing between molecules and because the arrangement of certain molecules in the plane of the plate of silica and gibbsite is hexagonal, it is possible for the gibbsite sheet to fit very closely onto the silica sheet and to be held there by relatively strong bonds. Thus, a possible clay mineral arrangement is that shown in Fig. 1–4b, in which a gibbsite sheet or a building block is bonded to a silica sheet. The building blocks shown represent layers of a unit molecular thickness in both cases. It is also possible, because of the hexagonal arrangement and the spacing of the molecules in the gibbsite and silica plates, for the gibbsite surface shown as *a-a* in Fig. 1–4*b* to be connected to the silica sheet shown as *b-b* by a chemical bond which is relatively weaker than that between the primary elements shown. If stacks of the basic building blocks are arranged in this way as shown in Fig. 1–4*c*, particles are produced of many molecular layers in thickness and with many times the basic intermolecular distance in lateral extent. The arrangement shown in Fig. 1–4*c* then gives rise to the clay mineral known as *kaolinite*. Typical kaolinite particles are from 0.1 to 10 microns in diameter and have a diameter-to-thickness ratio of the order of 40:1 or 50:1.

The hexagonal arrangement of the silica and gibbsite structures gives rise to a particle in which the 120° angles of the hexagon predominate. It is also observed that, in general, the particle does not simply consist of an hexagonal fragment with a plane upper and lower surface and relatively well-defined edges, but instead consists cf surfaces which are layered or terraced with many individual surfaces at various angles. This should be borne in mind when sketches such as those of Fig. 1–3 are drawn to illustrate the arrangement of particles. In such sketches it is usually shown for simplicity that the particles have flat surfaces.

Another possible arrangement of the basic building elements, gibbsite and silica, is shown in Fig. 1–5*a*, and it will be recognized that the bonds between the blocks at *a-a* and *b-b* in this diagram are both relatively strong. A mineral can be composed of aggregates of such elements by relatively weak bonding between the bases of adjacent silica sheet structures, as shown at *c-c* in Fig. 1–5*b*, and this structural arrangement is referred to as *montmorillonite*, an important class of clay minerals. Since the bond at *c-c* is extremely weak in montmorillonites, it forms a plane of cleavage (as does the *c-c* plane in Fig. 1–4*c* in kaolinites), and montmorillonites therefore tend to form clay particles with very small sizes. In fact, the surfaces of the silica tetrahedra at *c-c* in Fig. 1–5*b* frequently have greater affinity for other elements or molecules than for each other. This is particularly true in the case of the water molecule, so that montmoril-

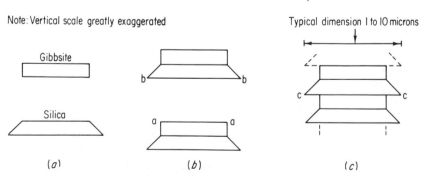

FIG. 1–4   *Structural arrangement in kaolinite clay particles.* (*a*) *Building blocks.* (*b*) *Structural units.* (*c*) *Kaolin particles.*

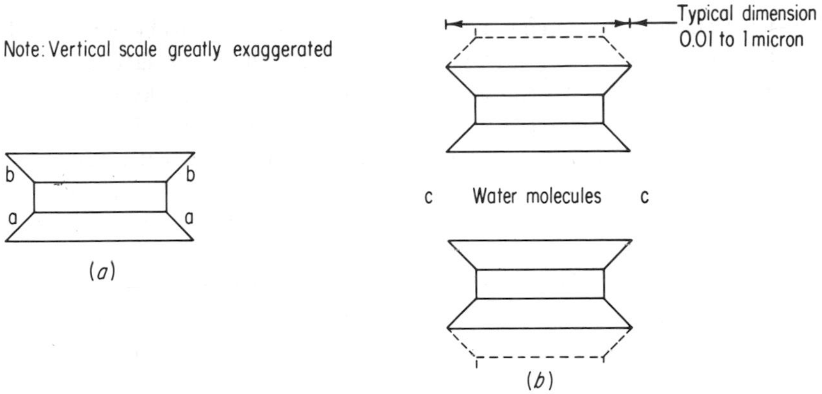

Note: Vertical scale greatly exaggerated

Typical dimension
0.01 to 1 micron

c   Water molecules   c

(a)

(b)

FIG. 1-5 *Structural arrangement in montmorillonite clay particles. (a) Structural unit. (b) Montmorillonite particle.*

lonite particle sheets tend to adsorb water in layers between or in the clay mineral atomic structure at the plane c-c of Fig. 1–5b. Thus, montmorillonite particles, besides being small, are capable of taking up large quantities of water (up to six molecular layers of water between each two molecular layers in the clay particles), and as a consequence the ratio of weight of water to weight of solids in a montmorillonite clay is usually high.

Typical montmorillonite clay particles range down to 0.1 micron in diameter, and they may have a diameter-to-thickness ratio of several hundred to one. Electron-microscope photographs of montmorillonite clay assemblages have been made; it is common in such pictures to find the individual montmorillonite particles difficult to identify because of their extremely small size. Indeed, in the early days of clay mineral analyses, before the techniques for examining the regularity of the crystalline structure of very small particles were available, clay minerals were thought to be amorphous in nature.

An element whose molecules are particularly suitable from the point of view of size and appropriate spacing for fitting into the space c-c between adjacent layers of the basic montmorillonite structure, as shown in Fig. 1–5b, is potassium, and it forms another clay mineral to which the name *illite* is given. The structure thus formed of alternate gibbsite and silica sheets with potassium molecules in between is relatively stable; it is shown schematically in Fig. 1–6.

It will be recognized that in all three of the clay mineral crystals there are different proportions of the various elements which go to make up the structure. Which clay mineral forms in a given environment depends on the ratios of the elements available in the aqueous solution out of which the minerals crystallize. It is seen, for example, that the aluminum-to-

Note: Vertical scale greatly exaggerated

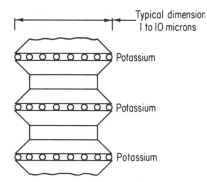

Fig. 1 – 6 *Structural arrangement in illite clay particles.*

silicon ratio is higher in a kaolinite than in a montmorillonite, so that conditions in which there is an abundance of aluminum or a scarcity of silicon are more suitable for the formation of kaolinites. Kaolinites frequently form from granite rocks under conditions of high temperature and pressure when water is available. Montmorillonites form from volcanic lavas or volcanic dust in the presence of water and heat. Thus, kaolinites are frequently associated with granite intrusions and montmorillonites with vulcanism. However, most clay minerals are moved from the site of their original formation by river and stream action and may be subjected to further alterations before deposition in a lake or sea bed.

The elements which make up the clay minerals are not restricted to the hydrogen, oxygen, silicon, and aluminum which have been mentioned previously, and other elements, such as iron and magnesium, may also have a place in, for example, the gibbsite structure and may replace some of the aluminum atoms in it. Consequently, within the basic structural arrangements found in Figs. 1–4 to 1–6, there is a great range and variety of clay minerals containing small amounts of other elements.

As pointed out earlier in this chapter, the clay minerals owe their existence to chemical processes and are themselves extremely small particles with high ratios of surface area to volume. Where clay minerals are concerned, their surface chemistry therefore plays an important role, and the behavior of clays is perhaps primarily dependent upon the nature and characteristics of the clay mineral surfaces. On these surfaces, foreign molecules or ions tend to be adsorbed. Since the aggregate behavior frequently depends on the movements at contacts between particles, the nature of the molecules present in the neighborhood of such contacts plays a part in determining the deformational or shearing strength properties of the soil.

When only small quantities of foreign molecules are present as, for example, in a clay soil whose pore water contains very few electrolytes, the clay particles tend to repel each other because of similar charges on the surface layers of molecules in adjacent particles. When the electrolytic concentration of the pore water is increased (as, for example, by adding salt to the pore water), some of the ions of the electrolyte arrange themselves near the clay mineral particle surfaces with a closeness and distribution of packing which depend on the chemical nature and size of the ions present. These ions modify the particle interactions to such an extent that at certain electrolytic concentrations the particles of clay are attracted toward one another and may, in fact, come in contact. It is generally thought that these contacts take place by the contact of the edge of one clay particle with the face of another, since the edges of clay particles are frequently electrically positive whereas their faces are electrically negative.

When the particles tend to repel each other in the presence of weak electrolytic solutions, the clay soil formed of such particles in water is referred to as *unaggregated* or *dispersed*. If sufficient electrolyte is added to cause the particles to come in contact with one another, the clay structure is referred to as *aggregated* or *flocculated*. The structural arrangement of dispersed and flocculated soils is shown in Fig. 1–7a and b, respectively, in which it should be observed that for convenience only a small range of particle sizes has been shown and the particle surfaces have generally been taken to be fairly smooth. In practice, as pointed out above, there will be a wide range of particle sizes present and the particles themselves will be very irregular. The arrangement of particles in flocculated clay soils as shown in Fig. 1–7b has been referred to as the "card-house structure."

The behavior of dispersed and aggregated clay soils composed of the same basic clay mineral particles varies widely, because in the first type of soil there is no contact between particles and in the second soil the particles form a structural interconnecting network. When one is carrying out a hydrometer test on a fine-grained soil and wishes to obtain the true distribution of particle sizes, it is desirable to obtain the soil in a dispersed condition so that the particles are truly separated and the test reflects the sizes of the individual particles. If for such a test a clay soil is made into a suspension with distilled water, it will frequently be found that several days must elapse before the small soil particles settle out. However, if a small quantity of sodium chloride, natural salt, or hydrochloric acid is

Fig. 1–7   *Structures in clay soils. (a) Dispersed. (b) Flocculated.*

Scale 1 to 50 microns, depending on clay type

(a)   (b)

added to the suspension at any time, the particles tend to aggregate or flocculate, and the flocculated clusters of particles settle much faster than the individual particles. The ratio of the external surface area to the mass of an aggregation of particles is smaller than that of an individual particle, and therefore the resistance of the cluster to fall through the water is smaller.

Because of the quantity of electrolytes present in natural waters in the soil, most soils in nature have been originally formed in a flocculated state by deposition in water. However, subsequent percolation and leaching of the original water from the soil may give rise to a structure which is in a potentially dispersed state, although the particles themselves have a flocculated structural arrangement indicative of the original mode of formation. Disturbance, remolding, or shearing of such soils leads to a disruption of the bonds or contacts between particles, and, because the new state is a potentially dispersed one, other bonds are not remade during the shearing process. It is therefore found that such soils decrease remarkably in shearing strength when they are disturbed or reworked.

Since the nature of the surfaces of the particles in a clay soil is so important, it follows that the mechanical behavior of a clay soil can be modified by the addition of chemicals to the soil, whereas, in general, the behavior of a sand or granular soil is not changed by the presence of chemicals other than binders or cements. For instance, if a sand composed of medium-sized particles is shaken up in water, the particles will settle out in the water at a rate which will be relatively independent of the amount of electrolyte in the water of the suspension. Also, if the strength of a sand soil in a certain state of density is measured by some means when the soil is saturated, it will be found that the strength does not depend on whether the water saturating the soil is fresh or salt. However, the behavior of a clay soil in a shearing test would depend very much on the electrolytic nature of the water which was mixed with the soil. Thus, when we take a sample of clay soil in the field, bring it into the laboratory, and mix it with either tap water or distilled water for the purpose of carrying out, for example, an Atterberg limit test on the remolded soil, the very nature of the soil may be changed by the character of the water which we add for the purpose of our test. Another laboratory, for example, might not obtain the same remolded soil properties if it were to use a different kind of water for mixing purposes.

The permeability of any given clay soil depends very much on the nature of the ions adsorbed on the surfaces of the clay particles (see Chapter 2, on permeability), and it is therefore possible to alter the degree of permeability to make the soil more or less impervious by adding another electrolyte to the pore water and either letting the water percolate the soil or mixing it with the soil. This technique is sometimes found of use in practice. Because of the importance of the physicochemical properties of clay particles, a great deal of effort has been devoted to the study of chemicals which may be employed to "stabilize" some soils by rendering them less permeable, stronger, or less compressible.[2]

1-3 DESCRIPTIVE PARAMETERS FOR SOILS

Every soil in nature or in the laboratory is composed of two or three material phases: the soil solids and the pore fluid, which may consist of gas, liquid, or both. In soils, the fluid phases are generally water and air. Therefore, convenient descriptive terms to indicate the condition of a particular soil by reference to the denseness or looseness of the granular material and the amount of gas or water present are convenient, and they will be established in this section.

We shall consider a soil element such as shown in Fig. 1–8, in which all of the grains of Fig. 1–8a are accumulated in Fig. 1–8b into a solid mass representing the bulk volume of the solid material present, the distributed water in the pores of the soil is collected in Fig. 1–8b into a single volume of liquid, and the gas bubbles into one volume of gas. It is conventional to indicate on the left-hand side of Fig. 1–8b terms describing the volumes of the various phases present and on the right-hand side of the figure terms representing the weights. Now we may proceed to derive ratios among the different parameters in order to describe the state of the soil. Because the weight of soil solids present in a given soil element is likely to remain constant throughout a given process (since we are rarely concerned with chemical processes in which the solids may be dissolved), it is convenient to consider the volume or weight of the solids to be a constant to which other variable quantities, such as the weight of water present in the element, may be referred. Thus, the first of our definitions concerns the ratio of the weight of water or weight of voids, since the weight of gas may generally be taken as zero, to the weight of soil. This is frequently referred to as the water content, in percent, and is given the

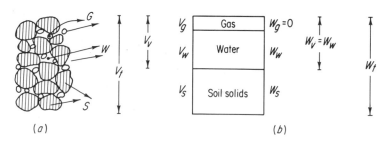

FIG. 1–8   *Soils as a three-phase system. (a) Soil mass. (b) Components of soil.*

symbol w. We define w by Eq. (1–1), in which, as in the subsequent equations, the terms are shown in Fig. 1–8b.

$$w = \frac{W_w}{W_s} \times 100 \tag{1-1}$$

The ratio of the volume of voids to that of the solids is also a convenient one to use. This is termed the *void ratio* and is given the symbol e, where e is given by the following equation:

$$e = \frac{V_v}{V_s} \tag{1-2}$$

As its name implies, e is a ratio of two numbers and is not a percent. Quite commonly in applications involving granular media other than soils a different measure of the volume of voids present is used. This is the *porosity* n, and it is defined to be the ratio of the volume of voids to the total volume of all phases together, as follows:

$$n = \frac{V_v}{V_t} \tag{1-3}$$

Commonly, n may be found as a percent rather than a ratio. A further item of interest concerns the amount of water actually present in the soil compared to the total pore volume available for it to occupy. This is referred to as the *degree of saturation* S, and it is defined by the equation

$$S = \frac{V_w}{V_v} \times 100 \tag{1-4}$$

Again, S is commonly employed as a percent. It is obvious that other relationships among the various quantities could be derived, but those given above are most commonly used.

The weights and volumes of water and soil solids present are related to each other by the specific gravities of the materials as, for example,

$$W_s = G_s \gamma_w V_s \tag{1-5}$$

where γ_w is the unit weight of water and G_s is the specific gravity of soil solids. It will be seen that various relationships can be obtained by substitution among the parameters described by Eqs. (1–1) to (1–4). One convenient relationship commonly used is given by Eq. (1–6):

$$G_s w = S e \tag{1-6}$$

We are also interested in the unit weight of the soil under various conditions. Perhaps the simplest or the most common is the total unit weight. This unit weight is usually represented by the symbol γ_t as defined by

$$\gamma_t = \frac{W_w + W_s}{V_t} \tag{1-7}$$

By the relations given in preceding equations it is obvious that γ_t can also be written in the forms

$$\gamma_t = (1 + w)\frac{W_s}{V_t} = (1 + w)(1 - n)\gamma_s = \frac{1 + w}{1 + e}\gamma_s \qquad (1\text{--}8)$$

If the whole soil element is submerged in water, we may be interested in the so-called *buoyant unit weight* γ_b of the material, and this is given by

$$\gamma_b = \gamma_t - \gamma_w \qquad (1\text{--}9)$$

which may also be expanded into a convenient form when the soil is completely saturated:

$$\gamma_b = \frac{G_s - 1}{1 + e}\gamma_w \qquad (1\text{--}10)$$

One final parameter which is frequently used in compaction experiments to determine the quantity of soil solids which has been rammed or compacted into unit total soil volume is the so-called *dry unit weight* γ_d of the material; this is given by the ratio of the weight of soil solids to the total volume of soil:

$$\gamma_d = \frac{W_s}{V_t} \qquad (1\text{--}11)$$

Convenient substitutions can be made for the terms on the right-hand side of Eq. (1–11) to give other forms of the equation in terms of previous parameters.

It is frequently of interest to determine the state of density of a soil as it occurs in nature or in a laboratory experiment with respect to the densest and loosest states in which the soil can exist. For granular soils, the term used to describe the soil's condition is the *relative density* R, which is expressed in terms of the void ratio of the soil in the state of interest e and the maximum and minimum void ratios which the soil can undergo, as follows:

$$R = \frac{e_{max} - e}{e_{max} - e_{min}} \times 100 \qquad (1\text{--}12)$$

Relative density is usually expressed as a percent, and it will be seen from Eq. (1–12) that, if the natural void ratio of the soil e is equal to the maximum void ratio the soil can attain, i.e., the soil is in its loosest state, the relative density of the material is 0 percent. If the void ratio e of the soil takes the smallest possible value, as it does when the material is in its densest state, the relative density of the soil is 100 percent. In practice, the minimum and maximum void ratios must be obtained for a given soil sample by means of laboratory tests and are subject to the methods employed for obtaining the given values, so that relative density is not a precise term to describe a given soil. The parameter which is used to describe the state of a clay soil and is analogous to the relative density is the *liquidity index* I_L of the soil, and this relates the existing water content of the material w to the water content of the clay and its liquid and plastic limits w_L and w_P, respectively, as follows:

$$I_L = \frac{w - w_P}{w_L - w_P} \times 100 \qquad (1\text{--}13)$$

Thus, if a soil exists in nature at a water content equal to the liquid limit, its liquidity index is 100 percent; if, on the other hand, it has a water content equal to its plastic limit, Eq. (1–13) indicates that its liquidity index is zero. Since the liquid and plastic limits are determined for remolded soils and since the shearing strength and deformational behavior of a soil depend on its existing structure, it is possible for a clay in nature to have a water content much higher than the liquid limit value without, in fact, the clay having zero shear strength; therefore, the liquidity index of a still-solid clay may, in fact, be much higher than 100 percent. Again, it may be possible to determine the water content of a clay in a dried-out state in nature and find that it is lower than the plastic limit, so that the liquidity index of the clay in this case would be negative.

PROBLEMS

1-1 A piece of clay taken from a sampling tube has a wet weight of 155.3 grams. After drying in an oven for 24 hr at 105°C its weight is 108.7 grams. What is the water content of the soil?

Answer   42.8 percent

1-2 If the clay of Prob. 1–1 has a liquid limit of 56.3 percent and a plastic limit of 22.5 percent, what is its liquidity index?

Answer   60 percent

1-3 The volume of the original piece of wet clay of Prob. 1–1 was measured and found to be 95.3 cm³. The specific gravity of the soil particles was 2.75. (a) What was the volume of soil solids present? (b) What was the volume of water present? (c) What was the degree of saturation of the original sample?

Answer   (a) 39.5 cm³, (b) 46.6 cm³, (c) 83.5 percent

1-4 Calculate the void ratio and porosity of the soil sample of the Probs. 1–1 to 1–3.   *Answer*   1.41, 0.586

1-5 Check that the values you have calculated in Prob. 1–4 satisfy Eq. (1–6).

1-6 What are the total, dry, and buoyant unit weights of the soil in Probs. 1–1 to 1–5?   *Answer*   117.7 pcf, 82.4 pcf, 55.3 pcf

1-7 The uniformity coefficient of a soil C_u is defined as the ratio of the 60 percent size to the 10 percent size. (The 60 percent size is that size of grain than which 60 percent of the soil by weight is finer; see Fig. 1–1b.) Calculate C_u for the four soils shown in Fig. 1–2.

1-8 Assign letter symbols to the soils of Figs. 1–1b and 1–2 by reference to Table 1–2.

1-9 What do you conclude about the classification of the soil of Probs. 1–1 and 1–2 by plotting its liquid limit and plasticity index on the plasticity chart in Table 1–2?

1-10 How do you expect the structures of clay soils deposited in fresh water and sea water to differ (see Fig. 1–7)?

1-11 Plot qualitatively on a diagram like Fig. 1–3 the volume changes you would expect of the two clays of Prob. 1–10 upon drying.

REFERENCES

1. R. F. LEGGET, "Geology and Engineering," 2d ed., McGraw-Hill Book Company, New York, 1962.
2. R. E. GRIM, "Applied Clay Mineralogy," McGraw-Hill Book Company, New York, 1962.

2 | Water in Soil

Water is present in all soils to some extent, and much of the science of soil mechanics is concerned with studies of the effect of the water on the behavior of the soil. The quantity of water present, the movement of the water through the soil, and also the chemical nature of the water are all of interest. The importance of the water in the soil in particular engineering circumstances depends on the type of soil involved. In general, in coarse-grained soils, the effect of water upon the soil strength or deformation characteristics is small, and engineers concern themselves primarily with the movement and level of water in such soils. The granular soils are also relatively incompressible, so that, as we shall see later, water flows through such soils largely without volume changes occurring in the soil; flow in this case is classified among problems of *steady-state seepage*, which will be studied in Chapter 4. In addition, since coarse-grained soils are composed of chemically stable minerals, as we saw in Chapter 1, the chemical nature of the pore water does not affect the state or condition of the soil during or after deposition except when salts crystallize out of the water to cement the particles weakly together.

On the other hand, in fine-grained materials the soil structure is itself compressible, and the movement of water through the soil is generally associated with volumetric changes in the material. Circumstances involving such volume changes accompanied by water flow are called *transient-flow* problems and are also studied in Chapter 4. It was pointed out in Chapter 1 that the chemical properties of the water in which fine soil grains are deposited determine the characteristic structure of the soil thereby formed and, as a consequence, influence its permeability, deformation, and strength properties. Variations in the chemical composition of the pore water during flow may also induce changes in the soil properties.

This chapter will be devoted to a study of the water present in soils and will define some of the parameters used to describe the state of the water.

2-1 PRESSURES AND HEADS

In the study of hydraulics a certain amount of energy is always associated with any element of water, and this energy for convenience is usually divided into (1) the energy in the water developed as a result of its position and state of compression, called its *potential energy*, and (2) the energy which the water element possesses as a result of its movement, termed *kinetic energy*. In any of the circumstances in which we study the movement of water in soils and the relation of water to soils in the present context, the velocities of water movement are always extremely small, so that, in relation to the potential energy, the kinetic energy of the water can always be neglected. Consequently, when we refer to the total energy of the water, we usually refer to the total potential energy, and this comprises two parts: (1) the energy which the water possesses as a result of its position and (2) the energy which it possesses as a result of its state of compression.

Since, in moving from one location to another, a unit mass of water can do work or take up work as a consequence of both its change in elevation with respect to some fixed datum and its change in volume as a result of change in pressure, it is usually convenient to express both energies or amounts of work in the same terms. Because it is easy to measure the total potential energy of water by examining the height to which it will rise in a standpipe inserted at the point under study, as shown in Fig. 2–1, the potential energy of water is usually referred to in terms of the height of the water column, or *head*, with respect to a fixed elevation. Thus, considering Fig. 2–1, which shows two standpipes inserted into soil through which water is flowing, it is seen that water rises in the first standpipe to a height greater than that in the second standpipe, with respect to the datum shown, by an amount $h_1 - h_2$. This means that if unit weight of water is transported from point 1 in the soil to point 2 in the soil, it does an amount of work represented by its weight times the head difference $h_1 - h_2$. Conversely, to move the unit weight of water from point 2 to point 1 would require the expenditure of the same amount of energy or work on the element of water. The conditions shown in Fig. 2–1 are intended to be static; that is to say, the water levels shown do not change with time. The two components of the energy can be examined as follows.

In general, an element of water at point 2, besides having a different elevation with respect to the datum from that of point 1, will be in a different state of compression or, alternatively expressed, will be subjected to a different hydrostatic pressure. To express the pressure in an element of water at a particular location conveniently, we consider the height of a

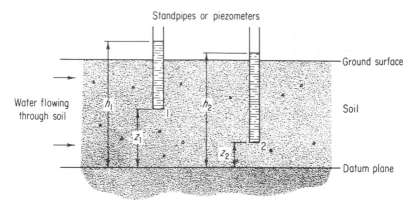

FIG. 2–1   *Total, elevation, and pressure heads in soil water.*

water column which would be required to exert that pressure. This height will be given by the pressure at the point divided by the unit weight of water when, as is usually the case in soil mechanics, the water temperature and unit weight do not differ substantially from place to place.

Now if the total energy possessed by any element of water is expressed in terms of head, we can say that the total head in the water is equal to the sum of the potential head measured above a fixed datum and the pressure head expressed, as above, in terms of the height of a water column. Expressing total head as h, elevation head as h_e, and pressure head as h_p, we can then write the following equation:

$$h = h_e + h_p \qquad\qquad (2\text{--}1)$$

If any two of these quantities are known in a given problem, the third can be calculated.

Referring to Fig. 2–1, the elevation head of the water in the soil at location 1 is given by z_1, the height of the water element above the fixed datum, and the pressure head is given by the distance $h_1 - z_1$ representing the head of water standing in the pipe required to maintain the pressure existing at point 1. At point 2 of Fig. 2–1 the elevation head is z_2 and the pressure head is $h_2 - z_2$. It will therefore be seen from Fig. 2–1 that the total head at point 1 is equal to the elevation head plus the pressure head, or $z_1 + (h_1 - z_1) = h_1$. Similarly, the total head at point 2 is equal to h_2. The actual pressure in the water at any point can always be obtained by multiplying the pressure head by the unit weight of water.

An illustration of the employment of heads is given in Fig. 2–2, which represents the flow of water through a horizontal tube filled with soil. At one end of the tube a standpipe is kept filled with water to a height of h_1 above a given datum, whereas at the other end of the tube another standpipe continuously overflows at the elevation h_2. Water is maintained at the two elevations h_1 and h_2 throughout an experiment. Considering that atmospheric pressure acts on all of the water in the system, we can refer the pressure or pressure head to atmospheric pressure as a datum. In this way, considering the point A at the top of the water in the standpipe to the left of the drawing, it will be seen that the pressure head at point A is zero (atmospheric), whereas the elevation head is h_1. At point B at the water surface in the tube at the right of the diagram the pressure head is again zero (atmospheric) and the elevation head in this case is h_2. Thus the total head at point A is h_1 (+0) and the total head at point B is h_2 (+0).

Considering point C at the center of one end of the soil-water interface, it will be seen that its elevation head is h_3 and its pressure head is equivalent to the column of water between it and point A, a vertical distance of $h_1 - h_3$. Adding the elevation and pressure heads at point C, we get a total head at point C of h_1. It is therefore seen that no head is lost between points A and C. This is true because in the present circumstances, although water is flowing, the velocity of flow is so small that there are essentially no energy or head losses as a result of friction between A and C, and consequently, to all intents and purposes, the measurable total head or energy in the water at point C is the same as that at point A.

Carrying out the same procedure for point D at the other end of the soil in Fig. 2–2, it will be seen that the total head there is h_2, the same as the total head at point B, since again the water loses no energy or head between points D and B. Consequently, we can see that, in flowing through the soil from point C to point D, the water has lost a total head of $h_1 - h_2$ as a result of performing frictional work on the soil between points C and D. Since, in this case, the elevation heads at C and D are the same, the head loss takes the form of a reduction in pressure head between C and D. Were the tube between C and D inclined either up or down with respect to the horizontal datum, it would be seen that in a more general case there would be a change of both elevation and pressure heads as the water flowed from one point to the other through the soil.

A particular case of this is shown in Fig. 2–3, in which the soil element is arranged vertically and the lower end of the tube corresponding to point B of Fig. 2–2 is kept submerged below the surface of water in a basin which continually overflows as a result of the flow of water through soil. The upper end of the tube corresponding to point A of Fig. 2–2 is maintained at a constant elevation by the addition of water to the tube. Thus,

Fig. 2–2 *Flow of water through a horizontal soil-filled tube.*

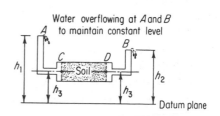

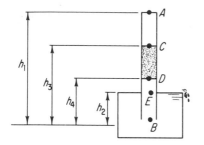

FIG. 2–3   *Heads in the case of vertical flow.*

water is flowing steadily through the soil between points C and D. In this case the total head at point A is once again h_1 measured from the arbitrary datum shown, and the total head at point B equals the total head at point E, or h_2. By adding together the elevation and pressure heads at point C it will be observed that the total head at C is equal to that at A, or h_1.

A similar calculation carried out for point D shows that the total head in the water below the base of the soil is everywhere the same and equal to h_2. Thus, as in Fig. 2–2, a total head loss of $h_1 - h_2$ occurs in the course of water flow through the soil; at point D, however, it will be seen that the pressure head is *lower* than it is at point E because point D is higher than E by the distance $h_4 - h_2$. Because point E is at the same level as the surface in the reservoir, it follows that the pressure at point E is zero (atmospheric), and consequently the pressure head at D must be below atmospheric or negative. Consequently, in flowing through the soil as shown in Fig. 2–3, the water is subjected to a change of both elevation and pressure heads.

It will be found that the elevation and pressure heads for other situations can also be determined most easily by studying, as above, the conditions on the continuous fluid or water side of the point being examined, rather than by trying to interpret, for example, the conditions at point D in Fig. 2–3 based on what is happening in the soil between points C and D. In a later section of this chapter the importance of the total head in the study of flow of water through soils will be explained.

2–2 CAPILLARITY, CAPILLARY RISE

If the situation shown in Fig. 2–3 is duplicated in Fig. 2–4a with the exception that, instead of a vertical tube containing soil, a tube with a fairly narrow bore is used, the common phenomenon of *capillary rise* may be observed. The water in the reservoir will rise up in the bore of the capillary tube to a height h_c above the level of water in the reservoir. The rise will stop at that point and no further movement of the water will take place. By establishing an arbitrary datum elevation as shown in Fig. 2–4a, so that the height of the water meniscus in the tube above the datum is h_1

and the height of the water in the reservoir is h_2, the conditions of total, elevation, and pressure heads can be examined.

After flow has ceased, there is no energy loss and consequently the total head everywhere in the water system must be the same. At point C the elevation head is h_2 and the pressure head is zero (atmospheric). At point B the elevation head is zero and the pressure head is h_2, so that the total head at both B and C is h_2. At point A in the water just under the meniscus the elevation head is h_1 and, since no work is being done by the system, the total head at point A must also be h_2, so that the pressure head at point A must be equal to $h_2 - h_1$, according to Eq. (2–1), or $-h_c$; that is to say, the pressure head at point A is below atmospheric. However, at point A' in the atmosphere just above the meniscus the pressure must obviously be atmospheric, and therefore a pressure difference exists across the meniscus.

The reason for this difference in pressure can be seen with reference to Fig. 2–4b, which shows the meniscus in greater detail. The atmospheric pressure (datum) is taken to be zero acting on the concave surface of the meniscus, and the pressure head, as has been seen, acting on the convex surface of the meniscus is equal to $-h_c$. We will take the surface tension per unit length of the meniscus perimeter at its intersection with the wall of the tube to be T_s. It can be seen that, in effect, the surface tension supports a column of fluid in the tube of height h_c; and by balancing the forces, we derive the following equation:

$$T_s \pi d \cos \alpha = \frac{\pi d^2}{4} h_c \gamma_w \tag{2-2}$$

from which

$$h_c = \frac{4 T_s \cos \alpha}{d \gamma_w} \tag{2-3}$$

Thus, for known values of surface tension, tube diameter d, and the contact angle α between the meniscus and the tube surface (an angle which depends on the molecular properties and affinities of the liquid for the material of the tube wall) the capillary rise can be calculated.

The situation with respect to capillary rise in soils is considerably more complicated than the foregoing by reason of the variations in pore space

FIG. 2–4 *Pressures and heads near meniscus.*

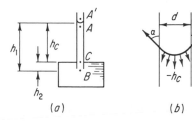

(a)

(b)

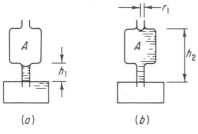

FIG. 2–5 *Capillary rise dependent on flow direction. (a) Rising level. (b) Falling level.*

size in even relatively uniform soils. In general, the distribution of pore size in a soil is not known, and it is not measured by any of the tests we commonly employ. However, the average pore diameter is frequently estimated to be approximately one-fifth the average grain diameter in a homogeneous soil and to be equivalent to the diameter of the 10 percent size in well-graded soils. It will be observed, however, that the diameter of any one pore will fluctuate considerably if one traces its passage through the soil. This variability of the pore diameters in soil causes the height of capillary rise in the soil to depend upon the direction of motion of the soil water. For example, in Fig. 2–5a water is rising in the capillary and will come to a stop at a height h_1 above the water table, where h_1 is dictated by the increasing radius of the capillary pore.

On the other hand, as shown in Fig. 2–5b, when the water table, initially at some greater height than shown, has been gradually lowered to the level shown, the pore space A will remain filled with water if the radius r_1 is sufficient to maintain the column of water of height h_2. Thus, in Fig. 2–5a and b, water will either fill or be absent from the pore space A for the same ultimate level of water table, depending on whether the water table is falling or rising. This reasoning applies to all soil pores, so that the capacity of a soil to store water depends upon its previous history of water table fluctuations. It is not possible to estimate the storage capacity of the soil pores for water by means of the porosity alone; it is necessary to define an *effective porosity*, which will vary with the history of water table movement.

In practice, in any soil, particularly a fine-grained soil as in the profile of Fig. 2–6, a water table is always associated with a capillary-rise zone in which the pressure is below atmospheric. For example, piezometers inserted in the soil as shown at A and B below the water table and in the capillary zone above it, respectively, will exhibit the true water table, although the level of water in tube B will be below its tip elevation, indicating a pressure below atmospheric. When water is flowing through soil,

FIG. 2–6 *Head in capillary-rise zone.*

it flows both in the capillary-rise zone and in the region below the true water table, but it is usually difficult to estimate the quantity of flow in the capillary region. Frequently, the soil in the capillary-rise zone is not saturated with water, but contains air spaces, and this may be also the case in the region below the true water table. So far we have referred to the water table or the "true" water table, and it can now be seen that these descriptions apply to the surface in a soil on which the water pressure is atmospheric. The surface is also called the *free surface* or, sometimes, the *phreatic surface*. Table 2–1 shows some values of capillary rise associated with various soils.

TABLE 2–1

Soil Type	Capillary-rise Range	
	cm	ft
Sand	10–100	$\frac{1}{2}$–3
Silt	100–1,000	3–30
Clay	>1,000	>30

To this point we have confined ourselves to the consideration of either the static water level in soil or the fact of the water flow through soil without concerning ourselves with the quantity of flow or any factors governing such flow. The next section will be devoted to the conditions surrounding the flow of water through soil.

2–3 DARCY'S LAW OF FLOW

Usually in engineering we represent flow conditions in terms of the pressure or the head acting on the fluid, and this is also convenient in soil studies. Consequently, it is desirable to discover the relationship between the pressure or head variations in the fluid in a soil and the velocity or quantity of flow of the fluid through the soil mass. We can determine this by performing an experiment of the type shown in Fig. 2–2. Let us consider in this figure that the soil occupies a container of cross-sectional area A, the soil mass has a length L, and at the ends of the container are attached two reservoirs both filled to and maintained at different levels. No air bubbles exist in the system. We maintain reservoirs A and B at heights h_1 and h_2, respectively, above the datum level. Since h_1 is greater than h_2, there will be a flow of water through the soil, and the quantity of flow may be measured. In an experiment we can maintain the level of B constant and increase the level of A while measuring the quantities of flow per unit time for each value of head in A. When we do this, we find that the relationship between flow and head difference across the soil is linear. From our previous discussion it is apparent that the heads h_1 and h_2 at reservoirs A and B measured as shown in Fig. 2–2 are total heads. Thus we conclude that, for the soil sample of given length and cross-sectional area in our example, the rate of flow through the soil is linearly proportional to the total head difference acting across the soil.

There are other variables which affect the flow rate. For example, if the length of the soil sample in Fig. 2–2 were doubled and the same experiment were carried out, it would be found that the quantity of flow had halved for each value of total head difference in the experiment. Alternatively, if the cross-sectional area of the soil in the container of Fig. 2–2 were doubled, twice the flow rate over the range of total head difference would be obtained. With this experimental evidence we may write an equation for the rate of water flow q through the soil in the following form:

$$q = kA \frac{h_1 - h_2}{L} \qquad (2\text{--}4)$$

where A and L are the cross-sectional area and length of the soil element, respectively, and h_1 and h_2 are the total heads acting across the ends of the sample perpendicular to the flow directions. The term k is the constant of proportionality relating the flow quantity to the other parameters.

If all factors were maintained constant and only the soil type changed, it would be found that the linearity of the expression would be unaltered for a wide range of soils. However, a fine-grained soil would give a small value of k in Eq. (2–4) and a coarse-grained soil a large value. The constant k is called the *coefficient of permeability* of a soil, or more commonly, the *permeability*, and has the dimensions of a velocity. A constant as defined in Eq. (2–4) includes in it the properties of viscosity and unit weight of the fluid flowing through the soil. The implication is convenient in soil engineering usage, since we concern ourselves almost exclusively with the flow of water alone, generally over a range of temperatures in which no significant variations in water viscosity or unit weight occur.

Attempts have been made to derive a general expression for the coefficient of permeability based on the size and shape of soil grains and pore spaces. We can see, however, that it is very difficult to describe a soil mathematically in these terms, and most mathematical derivations have referred to assemblages of tubes of constant but different cross sections. For a bundle of such tubes a flow law of the same form as Eq. (2–4) is obtained, but the numerical values for the constant which occurs in the mathematical derivation cannot easily be related to a given soil type. More recently, attempts have been to approach the problem of soil permeability from a statistical point of view. However, it is safe to say that all our further analytical work rests upon the empirical proof of Eq. (2–4) and the experimental determination of the value of the coefficient of permeability for each soil. In practice, when the problem of studying the flow of water through a region of soil arises, it is necessary to take samples of the soil in as undisturbed a condition as possible and to subject them to laboratory permeability tests arranged similarly to Figs. 2–2 and 2–3 for the purpose of obtaining values of the permeability coefficient in the different soils. Field permeability tests are also used.

The relationship given by Eq. (2–4) is known as Darcy's law after the French engineer who first obtained it, in the nineteenth century, in a testing arrangement similar to that shown in Fig. 2–3. The head difference, $h_1 - h_2$ in Eq. (2–4), and the length of the soil sample L are frequently combined in the form $(h_1 - h_2)/L$, which is referred to as the *total head gradient* or, more simply, as the *gradient* of water flow in the soil.

Darcy's law, Eq. (2–4), can be put into a form more convenient for mathematical operations. If we consider the soil element small enough that the total head difference across it can be written Δh and the length of the element Δx (or Δy or Δz), the gradient becomes $\Delta h/\Delta x$ (or $\Delta h/\Delta y$, etc.) and Darcy's law is written for flow in the x direction:

$$q_x = k_x A \frac{\Delta h}{\Delta x} \tag{2-5a}$$

In the limiting derivative form Eq. (2–5a) becomes

$$q_x = k_x A \frac{\partial h}{\partial x} \tag{2-5b}$$

If both sides of Eqs. (2–4) or (2–5) are divided by the area A, a velocity called the *superficial seepage velocity* v_x (v_y, etc.) is obtained:

$$v_x = k_x \frac{\partial h}{\partial x} \tag{2-6}$$

The velocity is described as superficial because it is obtained from the quantity of water flow divided by the total area of cross section including both soil and water. It is therefore not a velocity which could be measured in the pores, even as an average. The real average seepage velocity is obtained by dividing the superficial velocity by the porosity n, which represents, on the average, the ratio of pore space cross-sectional area to total area.

In Eqs. (2–5) and (2–6) we note that the coefficient of permeability (as well as the rate of flow and velocity) is given the subscript x (or y or z depending upon the direction considered); this notation is employed because most soils are stratified and possess permeabilities which differ in various directions. However, in most soils we can assume that the permeability in the x and y directions (both in the horizontal plane) are nearly equal to give $k_x = k_y$, which differ, usually to a considerable extent, from the permeability k_z in the vertical or z direction. The horizontal permeability of the soil is frequently many times greater than the vertical permeability, and in Chapter 4 we shall see how to take this difference into account in our analysis.

2-4 LIMITATIONS OF DARCY'S LAW

We now have an expression describing the rate of or a velocity of flow in soil in terms of a physical property of the medium and the total head gradient within it. The law depends on the fact that the grains of soil are relatively small and that the flow spaces between them are small. When soil is coarser than approximately a coarse sand size, the flow law has not the form of Eqs. (2–4) to (2–6), which are linear in the head difference, but must be so written that the velocity or quantity of flow depends on the head difference to some power. This nonlinearity can enter the behavior even in sands if the rate of flow (caused by head difference) is high enough. It is developed initially by the forces necessary to accelerate the water around the small particles (these forces are unimportant compared with the viscous drag of the soil particles at low velocities) and later, at still higher velocities, by a change in the type of flow in the soil from laminar to turbulent. Although this fact is known, the resulting nonlinear expression makes a mathematical solution for flow so difficult to obtain that very rarely have studies been made of the flow of water through a coarse soil.

On the other hand, in extremely fine-grained soils we have seen that an effective portion of the water may not be in the same physical state as normal water. When water flow takes place in such a fine soil as clay, it is not, therefore, surprising to find that Darcy's linear law also no longer strictly holds. While some empirical expressions have been suggested to describe the law of flow in clay, no general equations have been widely employed to date. Observations of flow in very dense or highly compacted clays indicate that no water flow at all takes place until the head gradient reaches some particular value which appears to depend on the soil and its state. In this respect "no flow" has reference to the limits of observation and discrimination in the very sensitive apparatus used in these experiments. In dense clay soils it has been found that limiting or threshold head gradients of approximately 30 to 60 (the gradient is, of course, dimensionless) must be applied before flow begins and that above this limiting gradient the flow is once again linear in the gradient.

We conclude this section on flow in soils with Table 2–2, which gives

TABLE 2-2

Soil Type	Grain-size Range, mm	Coefficient of Permeability	
		cm/sec	ft/yr
Gravel	>2	>1	>10^6
Sand	2–0.06	10^{-2}	10^4
Silt	0.06–0.002	10^{-6}	1
Clay	<0.002	<10^{-8}	<10^{-2}

approximate values of the coefficient of permeability for different granular materials.

PROBLEMS

2-1 In Fig. 2–2 the difference in water levels between points A and B is 1 ft and point A is 2.5 ft higher than points C and D. (a) What are the pressure heads at points C and D? (b) By how much do the total heads at points C and D differ?   *Answer* (a) 2.5 ft, 1.5 ft, (b) 1.0 ft

2-2 If the length of the soil sample in Fig. 2–2 is 2 ft, its cross-sectional area 0.3 ft^2, and the quantity of water flow through the soil is 3 in.^{3}/min, what is the permeability of the soil?   *Answer* 6,100 ft/yr

2-3 From the above permeability, what kind of soil do you guess is in the tube of Fig. 2–2?   *Answer* Fine sand

2-4 If the glass tube of Fig. 2–4a and b contains water and the walls are clean, the angle α is 0°. The surface tension of water is 0.078 gm/cm. Calculate the tube diameters which will give the capillary-rise figures of 10, 100, and 1,000 cm in Table 2–1.   *Answer* 0.31, 0.031, 0.0031 mm

2-5 Do the tube diameters you get in Prob. 2–4 appear to be the same sizes as you would expect the pores of the soil to be in Table 2–1? Compare them with the grain sizes given for those soils by the MIT size system of Table 1–1.

Answer Grain sizes are: coarse sand, 2 to 0.6 mm; coarse silt, 0.06 to 0.02 mm; fine silt, 0.006 to 0.002 mm. With a mean pore diameter about one-fifth of the grain diameter it appears that the first tube size (0.31 mm) of Prob. 2–4 is about the same size as the mean pore diameter of a coarse sand; the second and third tube sizes are a little larger than the mean pore diameters of a coarse silt and fine silt, respectively.

2-6 For a capillary rise of 1 ft (a) what is the pressure head in the water immediately below the meniscus in a tube, as at point A in Fig. 2–4?

(b) How does the total head at point A compare with the total heads at points B and C?

        Answer   (a) 1 ft *below* atmospheric pressure. (b) The same.

2-7 If the water table slopes from left to right in Fig. 2–6 so that water is flowing from B to A and the height of capillary rise is constant, is water flowing through the capillary zone too?

        Answer   Yes, since the total head at a point in the capillary zone such as B is greater than that at an equivalent point in the capillary zone above point A.

2-8 Do you know any other flow laws which have the same form as Darcy's law, Eq. (2–4), in which a flow amount is linearly proportional to the gradient of some quantity?

        Answer   Ohm's law in electricity: electrical current is proportional to voltage gradient. Fourier law of heat conduction: heat flow is proportional to temperature gradient.

3 | Stresses Acting in Soils

Much of soil mechanics is concerned with the response of soils to the stresses developed in them by loads caused either by footings or foundations at ground surface, by piles, or by the construction of tunnels, shafts, and trenches below ground. It is desirable, in practice, to know something about two differing classes of soil behavior under the application of stresses: the first concerns the amount of deformation which will occur and will lead to the settlement or distortion of structures constructed on or in the soil. The second class relates to the maximum value of the load or stresses which can be imposed on the ground before the soil fails or yields. Although the second effect is the more important from the point of view of the safety or stability of a structure, the first is also significant, since unequal settlements of different parts of a structure can lead to the cracking of walls, dislocation of utilities, and to a generally unsightly appearance. Excessive settlement may even impair the structural stability of the structure before the soil itself has been stressed to failure.

Both deformation and failure can be considered part of one logical sequence of loading. In Fig. 3–1 the horizontal axis shows the load, force, or stress applied to, say, a structure resting on soil, and the vertical axis gives the settlement or displacement of the structure. Curve A of this figure shows that a definite amount of movement or displacement accompanies each increase in load up to a certain load value at which unlimited movement occurs. This last condition is called *yield, rupture,* or *failure;* in Fig. 3–1 the asymptotic value of load at which it occurs on curve A is called F_u. If we are considering a footing placed on soil, for example, the load or stress F_u is called the *bearing capacity* of the soil or footing.

It is well to recognize at this stage that, for all soils, any curve

43

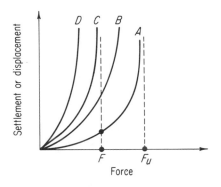

FIG. 3–1   *Relationship between applied force and displacement.*

plotted as shown in Fig. 3–1 is obtained from a test in which the length of time for which the soil is stressed, or the rate at which the load or force is applied to the ground, can be varied. We find, in practice, that, unlike the response of steel, the response of the soil depends on the rate at which we stress it. Thus, the other curves of Fig. 3–1, B, C, and D, might represent the load-displacement behavior of the same soil at different loading rates. In this illustration, curves A to D represent rates progressively *slower* by orders of magnitude. Alternatively, a constant load F applied at the ground surface by means of, say, a footing, will not produce one value of settlement or displacement, but a settlement which gradually increases in time as shown in Fig. 3–1, where the time interval corresponds successively to the curves of different loading rates.

Similarly, the failure load for one structure and soil is not, in general, constant, but can also be considered to vary with time. For example, were the force F in Fig. 3–1 to be applied for a very long time, we might find that the settlement or displacement, increasing in time, eventually would reach excessive values for the particular structure under consideration, since F is the asymptotic value of curve C. We might therefore consider that failure had occurred at this value F of the load, with a time delay, whereas the structure would be considered quite safe for short-time loadings. This is one aspect of soil behavior which makes analyses and design of soil structures subject to uncertainty and which makes it difficult to predict the behavior of soil under given circumstances with accuracy.

It is generally considered that a curve exists, such as D in Fig. 1, which represents the limiting behavior of the system under all rates of loading smaller than some low value. If this is true, then when the load applied is smaller than the asymptotic load of curve D, it may be concluded that failure will *never* take place if all other conditions are unaltered.

Before going on to develop more quantitative descriptions of soil behavior than given by the curves of Fig. 3–1, we must consider an element of soil in equilibrium under the system of forces acting on it.

3–1 EQUILIBRIUM OF SOIL ELEMENT

Any analysis begins by establishing a set of symbols and a convention which refers them to a system of axes. In Fig. 3–2a elements of material touch each other under the action of normal stresses σ and shearing stresses τ. In soil mechanics a normal stress σ acting at right angles to a given face is considered to be positive when it is acting in a direction tending to compress the element. (Frequently, in the study of other materials, such as steel, such stresses are considered to be positive when they extend the element.) The normal stresses acting parallel to the x, y and z axes are given the subscripts x, y, and z and become σ_x, σ_y, and σ_z, respectively; we term the faces of a material element perpendicular to the x, y, and z axes the x, y, and z faces, respectively.

Consequently, the shearing stresses acting on the different faces of the prismatic element can be described by means of two subscripts, the first describing the face on which the shearing stress is acting, and the second indicating the direction of the stress, so that τ_{zx} indicates a shearing stress directed parallel to the x axis acting on a face perpendicular to the z axis. When one looks in the positive direction of an axis, a shearing stress acting on a viewed face of an element is positive when it acts in the positive direction of one of the other axes. Viewed in the opposite direction a positive shearing stress on the face seen acts in the negative direction of one of the other axes. Figure 3–2a shows this with respect to the surface across which shearing stresses are transmitted in an element.

The stresses acting on the opposite pairs of faces of any soil element in a generally stressed region will be different. If we take the size of the element to be small, we can consider that the stresses vary linearly with distance to a close enough approximation. Then, with σ_x representing the normal stress at the center of the element in Fig. 3–2b, the normal stresses

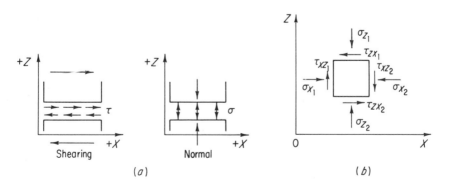

Fig. 3–2   *Stresses acting on an element. (a) Positive stresses. (b) Equilibrium of element.*

on the two opposite x faces become

$$\left.\begin{array}{c} \sigma_{x_1} \\ \sigma_{x_2} \end{array}\right\} = \sigma_x \mp \frac{\Delta\sigma_x}{2} \tag{3-1}$$

where $\Delta\sigma_x$ is the change in σ_x across the element. Similar expressions can be written for all the other stress components in terms of their values at the element's center, σ_z, τ_{zz}, and τ_{zx}.

The static equilibrium of the element in Fig. 3–2b is now to be examined. Taking moments about the center of the element, we find that the normal stresses on the faces have no moments if we assume again that the element is small enough that all stresses are uniform over the faces on which they act. The only contribution to the moments comes from the shearing stresses, whose moments must sum to zero if the element is not in motion, so that we get

$$\left[\left(\tau_{xz} - \frac{\Delta\tau_{xz}}{2}\right) + \left(\tau_{xz} + \frac{\Delta\tau_{xz}}{2}\right)\right]\Delta z \frac{\Delta x}{2}$$
$$- \left[\left(\tau_{zx} - \frac{\Delta\tau_{zx}}{2}\right) + \left(\tau_{zx} + \frac{\Delta\tau_{zx}}{2}\right)\right]\Delta x \frac{\Delta z}{2} = 0 \tag{3-2}$$

We conclude from Eq. (3–2) that $\tau_{xz} = \tau_{zx}$ and that, in the plane stress state shown in Fig. 3–2b, there are only three independent stresses at a point: σ_x, σ_z, and τ_{zz}. If the analysis were carried out in three dimensions, six independent stresses σ_x, σ_y, σ_z, τ_{xy}, τ_{xz}, τ_{yz} would result.

To proceed with the examination of the static equilibrium of the element of Fig. 3–2b, we resolve the forces acting on the element in the x direction to obtain

$$\left[\left(\sigma_x - \frac{\Delta\sigma_x}{2}\right) - \left(\sigma_x + \frac{\Delta\sigma_x}{2}\right)\right]\Delta z$$
$$+ \left[\left(\tau_{xz} - \frac{\Delta\tau_{xz}}{2}\right) - \left(\tau_{xz} + \frac{\Delta\tau_{xz}}{2}\right)\right]\Delta x = 0$$

or, dividing throughout by $\Delta x\,\Delta z$,

$$\frac{\Delta\sigma_x}{\Delta x} + \frac{\Delta\tau_{xz}}{\Delta z} = 0 \tag{3-3a}$$

Were the element accelerating under the influence of the applied forces, the right-hand side of the equation would not be equal to zero but would include the mass of the element times its component of acceleration in the x direction.

Resolving the forces in the z direction in the same way, we get the equation

$$\frac{\Delta\sigma_z}{\Delta z} + \frac{\Delta\tau_{xz}}{\Delta x} + \gamma_t = 0 \tag{3-3b}$$

in which the third term arises from consideration of the weight of the element. Here we have assumed, as is conventional, that the z axis is aligned parallel to the direction of gravity.

The two equations (3–3a) and (3–3b) are the equations of equilibrium in two dimensions; they may be written in the partial-differential form

$$\frac{\partial \sigma_x}{\partial x} + \frac{\partial \tau_{xz}}{\partial z} = 0 \tag{3-4a}$$

$$\frac{\partial \sigma_z}{\partial z} + \frac{\partial \tau_{xz}}{\partial x} + \gamma_t = 0 \tag{3-4b}$$

In three dimensions, these equations are modified by the addition of terms including the σ_y, σ_{xy}, and τ_{yz} stresses and another equation is added by considering the equilibrium of the element in the y direction. The equations may also be expressed in radial or polar coordinates, which are more convenient for some problems.

3 2 MOHR'S CIRCLE OF STRESS

Sometimes in soil mechanics problems the stresses acting on the faces of an element such as that shown in Fig. 3–2b are known and the engineer needs to know the stresses on some other plane or face at a given angle to the axes. This situation is presented in Fig. 3–3a. Here we are given the stresses σ_x, σ_z, and τ_{xz} acting on the faces AB and BC of an elemental volume of material and we would like to know the stresses σ and τ acting on the face AC where AC makes the angle θ with BC. These can be obtained by resolving the forces acting on the element in directions perpendicular to and parallel to AC. Carrying this out we get, resolving perpendicularly to AC,

$$\sigma \overline{AC} = \sigma_x \overline{AB} \sin \theta + \sigma_z \overline{BC} \cos \theta + \tau_{xz}(\overline{AB} \cos \theta + \overline{BC} \sin \theta)$$

or, dividing by $\overline{AC}$ throughout,

$$\sigma = \sigma_x \sin^2\theta + \sigma_z \cos^2\theta + 2\tau_{xz} \sin \theta \cos \theta \tag{3-5}$$

Resolving parallel to AC gives

$$\tau = \frac{\sigma_z - \sigma_x}{2} \sin 2\theta - \tau_{xz} \cos 2\theta \tag{3-6}$$

For constant values of σ_x, σ_z, and τ_{xz}, Eqs. (3–5) and (3–6) can be evaluated to give σ and τ as functions of θ. The result is shown plotted in Fig. 3–3b. It can be seen that the maximum and minimum values of the normal stresses occur at values of θ for which τ is zero. These angles are given by making the right-hand side of Eq. (3–6) equal to zero, to get

$$\tan 2\theta = \frac{2\tau_{xz}}{\sigma_z - \sigma_x} \tag{3-7}$$

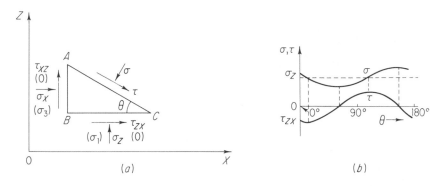

FIG. 3-3   *Stresses on arbitrary plane through element.* (a) *Element in equilibrium.* (b) *Stress variation with angle.*

The largest and smallest magnitudes of the normal stresses are called the *principal stresses* and are conventionally given the symbols σ_1 and σ_3, respectively. By substitution of Eq. (3–7) into Eq. (3–5), these values are

$$\left.\begin{array}{c}\sigma_1\\\sigma_3\end{array}\right\} = \frac{\sigma_x + \sigma_z}{2} \pm \left(\frac{\sigma_z - \sigma_x}{2}\right)^2 + \tau_{xz}{}^2 \tag{3-8}$$

It can be seen from Eq. (3–7) that the zero values of shearing stress occur in planes at angles of 90° to one another. These are the surfaces on which the principal stresses act and are called *principal planes*. The mean value of the normal stresses, shown by the dashed line on Fig. 3–3b, is $(\sigma_x + \sigma_z)/2$.

If, in Fig. 3–3a, we describe the plane AC to lie at an angle α to one of the principal planes instead of making the angle θ to an arbitrary surface BC on which both normal and shearing stresses act, Eqs. (3–5) and (3–6) become

$$\sigma = \sigma_3 \sin^2\alpha + \sigma_1 \cos^2\alpha$$

or

$$\sigma = \frac{\sigma_1 + \sigma_3}{2} + \frac{\sigma_1 - \sigma_3}{2} \cos 2\alpha \tag{3-9}$$

and

$$\tau = \frac{\sigma_1 - \sigma_3}{2} \sin 2\alpha \tag{3-10}$$

respectively, where σ_z and σ_x of Eqs. (3–5) and (3–6) are replaced by σ_1 and σ_3 and τ_{xz} is made equal to zero.

Equations (3–9) and (3–10) describe a circle, as shown in Fig. 3–4, where the coordinate axes have been chosen to be σ and τ. The diameter of the circle is $\sigma_1 - \sigma_3$, and its center is distant $(\sigma_1 + \sigma_3)/2$ from the

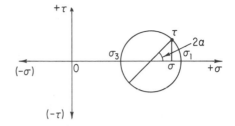

FIG. 3-4   *Mohr diagram of stresses.*

origin of the coordinates; the circle cuts the σ axis at two points whose coordinates represent the principal stresses $(\sigma_1,0)$ and $(\sigma_3,0)$. This construction is called the Mohr diagram. Since the circle in the diagram was developed from the consideration of the equilibrium of an element, the stress condition on any plane through a point in a material at equilibrium under the stresses applied to it is represented by a point on the Mohr circle. The circle can be constructed if the stresses σ_x, σ_z, and τ_{xz} are known at a point or if the principal stresses have been determined. Once the circle has been drawn, say, from the principal stresses acting on planes of known direction, the stresses on another plane at a given angle α to the major principal plane can be obtained from the intersection of the ray from the center of the circle, at the angle 2α to the σ axis, with the circumference of the circle. If the circle was drawn from a knowledge of the σ_x, σ_z, and τ_{xz} stresses, a similar construction will give the stresses on other planes. The Mohr circle will be referred to again when the shearing strength of soils is being studied.

If we were to carry out a similar but more complicated analysis analogous to that shown in Fig. 3-3 but involving three dimensions, we would find that there are three principal planes and three principal stresses (given the symbols σ_1, σ_2, and σ_3) in the three-dimensional stress state. In this circumstance the three principal stresses represent major, intermediate, and minor normal stresses, respectively. It should be noticed that, up to this stage in our discussion, no material behavior has been taken into account and no approximations have been made, so that the existence in general of three principal planes and stresses does not depend upon the properties of the material that we are studying. In other words, we have examined only a stress system in equilibrium, and we have not said that the material must be uniform or isotropic or possess certain stress-deformation relationships. The principal stresses therefore have general validity for all materials and all conditions of deformation.

The equations of equilibrium are the starting point of analyses in problems of applied mechanics; to proceed further we require a knowledge of the properties of the material under study. It is now appropriate to consider the relationship of the deformations that develop in materials to the stresses that cause them.

3-3 DEFORMATIONS IN MATERIALS

It is tacitly assumed that the application of stress to any material will result in a deformation or distortion of the material. If no deformation or distortion occurs, the substance is termed *rigid*. If, under the application of stress, deformations or distortions of material take place, perhaps changing in time but eventually coming to a stop while the stress state is maintained, then the material is referred to as a *solid*. If, under the continued application of a stress state, the deformations or distortions increase without limit, the material is said to be a *fluid*.

We distinguish between the gaseous and liquid states of fluids. In the gaseous state the molecules are separated by relatively large distances so that intermolecular forces play almost no part in the fluid characteristics, except at high pressures, whereas in the liquid state the molecules of the material are close enough together that intermolecular forces determine the material deformational characteristics. The distances between molecules in liquids are comparable to those in solids. However, the molecules in liquids are mobile and no molecular arrangement persists for more than a short time; on the other hand, the arrangement of molecules in a solid is constant and, in crystalline solids, is regular. Molecules in a solid change their position either not at all or very slowly, whereas molecules in a liquid are in a state of continuous motion.

It is now appropriate to consider the positions of points and the lengths and directions of line elements in a material before and after deformation arising from any cause whatever. Certain geometrical relations among the various quantities are of importance for subsequent work and will be developed below.

In Fig. 3–5 a rectangular element $OABC$ of original size Δx by Δz is displaced and changed in shape and size by the application of stresses to the material of which it is a component. Point O is displaced distances u in the x direction and w in the z direction, but we choose, for convenience, to bring the displaced point O back to its original position so that the change in size and shape of $OABC$ can be more easily visual-

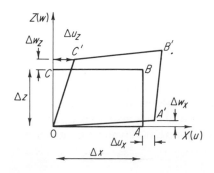

Fig. 3–5 *Displacements and strains.*

ized. We consider that the points A, B, and C move to positions A', B', and C' and that the increases of the u and w displacements in the x direction at point A are Δu_x and Δw_x, respectively. At point C the increases of the u and w displacements in the z direction are Δu_z and Δw_z. In Fig. 3–5 the lengths of all the boundaries of the element and its volume increase; the conclusions we draw would be unchanged if a contraction were to take place.

If all the extensions are small, then the increase in length of OA is very nearly equal to Δu_x and that of OC to Δw_z. There are a number of ways in which we may describe these changes in length, and we shall choose to do so by defining the *strain* of an elemental length such as Δx or Δz to be the ratio of the increase in length to the original length. Thus, we have

$$\epsilon_x = \frac{\Delta u_x}{\Delta x} \qquad \epsilon_z = \frac{\Delta w_z}{\Delta z} \tag{3–11}$$

where ϵ_x and ϵ_z are the strains in the x and z directions, respectively.

In differential form the strains are represented by the partial derivatives of the displacements in the x and z directions

$$\epsilon_x = \frac{\partial u}{\partial x} \qquad \epsilon_z = \frac{\partial w}{\partial z} \tag{3–12}$$

We notice further in Fig. 3–5 that the original right angle AOC has decreased as a result of the displacements the element has undergone. The decrease is equal to the sum of the two small angles $A'OA$ and $C'OC$. If we express these angles in radians, we see that the total angular change in radians, which is referred to as the *shearing strain* ϕ_{xz}, can be given to a close approximation by the equation

$$\phi_{xz} = \frac{\Delta w_x}{\Delta x} + \frac{\Delta u_z}{\Delta z} \tag{3–13}$$

In derivatives, this becomes

$$\phi_{xz} = \frac{\partial w}{\partial x} + \frac{\partial u}{\partial z} \tag{3–14}$$

It will be observed that, in three dimensions, a further normal strain ϵ_y and two more shearing strains, ϕ_{xy} and ϕ_{yz}, can be derived to describe the changes in length and angles which occur.

We now wish to point out that two variables u and w, in two-dimensional space, are sufficient to describe the displacement of a point, but that we have also chosen to describe the changes in length and angles of any element by three strain variables. As a consequence, there must be a relationship between the three strains to limit their relative magnitudes. This is the *compatibility equation*, or equation of consistent deformation, and it will be given in differential form only; it is obtained by differentiating the first of Eqs. (3–12) twice with respect to z, the second of Eqs. (3–12) twice with respect to x, and Eq. (3–14) once each with respect to x and z. Finally, the differentiated forms of Eqs. (3–12) are substituted into the differentiated version of Eq. (3–14) to give

$$\frac{\partial^2 \epsilon_x}{\partial z^2} + \frac{\partial^2 \epsilon_z}{\partial x^2} + \frac{\partial^2 \phi_{xz}}{\partial x\, \partial z} = 0 \tag{3-15}$$

In three dimensions, there are five more compatibility equations. When a problem is more suitably described by a radial or polar coordinate system, appropriate radial and angular strains may be defined in terms of radial and angular displacements by equations similar to Eqs. (3–11) to (3–14). The compatibility relations can also be obtained in a different form for these situations.

3-4   CONCEPTS OF MATERIAL BEHAVIOR

If, under the application of an applied stress, deformation takes place in a solid relatively quickly and comes to a stop at a finite amount of deformation or distortion for a given stress state relatively quickly and the material regains its initial shape and dimensions upon removal of the stress state, the material is referred to as *elastic.*

If, in the elastic solid, the amount of deformation, when this is properly defined, is directly proportional to the applied stresses, the material is described as a *linearly elastic solid.* If the deformation is not proportional to the applied stresses, the material is referred to as a *nonlinearly elastic solid.* The mechanics and mathematics associated with nonlinearly elastic materials are very complex, and they will not be discussed further in this book.

The characteristic behavior of *liquids* under stresses is described by a relationship between the *rate* (not *amount,* as in solids) of deformation and the applied stresses. If the rate of deformation of a liquid under the applied-stress system is linearly proportional to the stresses, the liquid is said to behave in a *linearly viscous* manner and is also frequently described as a *Newtonian liquid,* since Newton first described this behavior. Should the rate of deformation not be linearly proportional to the applied-stress system, the material is generally described as nonlinearly viscous or as being non-Newtonian in its behavior.

In some solids, notably among the polymer materials, the material deforms slowly under applied stress but eventually reaches an equilibrium shape; on removal of stress, recovery also occurs slowly until eventually the original shape and dimensions are again attained. These materials are called *viscoelastic solids.*

We have now reached a stage at which we must consider the quantitative relations between the stresses applied to materials and the deformations and distortions experienced by the substance under stress.

3-5 STRESS–STRAIN RELATIONS; LINEAR ELASTICITY

It is experimentally observed with many materials that, in a compression test as shown in Fig. 3–6, the application of the compressive stress σ_1 on the ends of the specimen (it is a principal stress, since we are applying no shearing stress to these faces) results in a small diminution of the specimen's length by ΔL which, by the definition of Eq. (3–11), is equal to $\epsilon_1(L_0)$, and it is further noticed that this deformation is accompanied by a lateral expansion of the specimen by an amount ΔB which is, by Eq. (3–11), equal to $\epsilon_3(B_0)$. It is conventional to take the strain ϵ_2 and stress σ_2 to be occurring in a direction perpendicular to the plane of the paper. On removal of stress the specimen returns to its original dimensions, and is therefore elastic.[1] If the change in length is linearly proportional to the stress, the substance is linearly elastic and the relationship between strain and stress can be described by a constant E, which is usually called Young's modulus, according to the equation

$$\sigma_1 = E\epsilon_1$$

or

$$\epsilon_1 = \frac{\sigma_1}{E} \tag{3-16}$$

If the material behavior can be described in this way to a close enough approximation, the lateral expansion of the sample may also be linearly

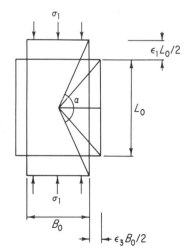

Fig. 3–6   *Normal strains.*

related to the vertical contraction, so that

$$\epsilon_3 = - \nu\epsilon_1$$

or, by Eq. (3–16),

$$\epsilon_3 = - \frac{\nu\sigma_1}{E} \tag{3-17}$$

where ν, another proportionality constant, is called *Poisson's ratio*. Now let us imagine that in Fig. 3–6 the stress σ_1 is removed and, instead, a stress σ_3 is applied to the lateral faces of the specimen. Once again, the strain ϵ_3 in many materials may be observed to be proportional to the stress σ_3, but we may distinguish two possibilities: (1) the constant of proportionality is the same as in the first case, as in Eq. (3–16), or (2) the constant is different. When (1) is true and also holds for the strains ϵ_3 produced by stress σ_3, the material is called *isotropic*. If the material properties differ in various directions as in (2), the substance is said to be *anisotropic*. In nature, soils are generally anisotropic, since their properties in horizontal directions are usually different from those in the vertical direction. However, they are frequently assumed to be isotropic, since the mathematical solutions for problems of stress distribution in anisotropic materials are difficult to obtain. Also, it is not easy to measure soil properties in the horizontal direction effectively.

If soil properties such as Young's modulus and permeability are the same at all points throughout a soil mass, the mass is said to be *homogeneous;* the material may be isotropic or anisotropic. In real soils, layers of different properties usually exist, so that the soil properties vary from point to point in a region of soils, and in this case the region is described as *nonhomogeneous* or *inhomogeneous*.

Should the specimen shown in Fig. 3–6 be subjected to three different normal stresses at once, the resulting small strains may all be super-imposed (added together) as if the stresses acted independently of one another. This is an important aspect of the behavior of linearly elastic materials. Thus, for example,

$$\epsilon_1 = \frac{1}{E} [\sigma_1 - \nu(\sigma_2 + \sigma_3)] \tag{3-18a}$$

and

$$\epsilon_2 = \frac{1}{E} [\sigma_2 - \nu(\sigma_1 + \sigma_3)] \tag{3-18b}$$

$$\epsilon_3 = \frac{1}{E} [\sigma_3 - \nu(\sigma_2 + \sigma_1)] \tag{3-18c}$$

If the material is isotropic, or if the axes of anisotropy are at right angles to each other and are coincident with the principal axes, there are no shearing strains in the planes on which the principal stresses act. These planes are therefore planes of principal strain, and the strains of Eqs. (3–18) are correctly shown to be principal strains ϵ_1, ϵ_2, and ϵ_3. The original right angles of the prism shown in Fig. 3–6 remain right angles after the application of the stress σ_1. However, shearing strains do occur on other planes, as may be seen by considering the change in the angle α, for example, in Fig. 3–6.

When a prism of material is stressed by three principal stresses which have the same value, equal, say, to σ_h, the normal strains in an isotropic material are also equal and, from Eqs. (3–18), are given by

$$\epsilon_h = \epsilon_1 = \epsilon_2 = \epsilon_3 = \frac{1 - 2\nu}{E} \sigma_h \tag{3-19}$$

Since all the strains are equal and positive, it will be recognized from the geometry of strain, as shown in Fig. 3–7, that there are in this case no shearing strains on any plane through a material element. This is called *hydrostatic distribution* of pressure or stress, since it is the state of stress in fluids at rest. These strains are not equal in this case if the substance is anisotropic, and, consequently, shearing strains occur in anisotropic materials under hydrostatic pressures.

To consider the volume change of an element such as that shown in Fig. 3–6 under three principal stresses σ_1, σ_2, and σ_3, when the element dimension in the direction perpendicular to the paper is D_0, we first obtain the original volume V_0:

$$V_0 = L_0 B_0 D_0 \tag{3-20}$$

The new volume V will be

$$V = (1 - \epsilon_1) L_0 (1 - \epsilon_3) B_0 (1 - \epsilon_2) D_0$$

or

$$V = (1 - \epsilon_1)(1 - \epsilon_2)(1 - \epsilon_3) V_0 \tag{3-21}$$

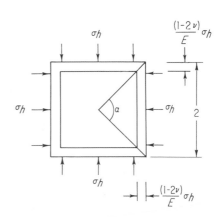

Fig. 3–7 *Hydrostatic stress state.*

If we define a volume strain d in the same way as the normal strains were defined,

$$d = \frac{V_0 - V}{V_0} \tag{3-22}$$

we get by substituting Eq. (3–21) in Eq. (3–22)

$$d = 1 - (1 - \epsilon_1)(1 - \epsilon_2)(1 - \epsilon_3) \tag{3-23}$$

In multiplying out the right-hand side of Eq. (3–23) double and triple products of the strains appear; but the strains are taken to be very small in our development here, so we can consider that products of strains are negligible with respect to the strains themselves. We can therefore approximate Eq. (3–23) by the expression

$$d = \epsilon_1 + \epsilon_2 + \epsilon_3 \tag{3-24}$$

In the special case of hydrostatic compression we find from Eqs. (3–19) and (3–24) that

$$d = \frac{3(1 - 2\nu)}{E} \sigma_h \tag{3-25}$$

Frequently this is further abbreviated by defining a *bulk modulus K*, where

$$K = \frac{E}{3(1 - 2\nu)} \tag{3-26}$$

so that

$$d = \frac{\sigma_h}{K} \tag{3-27}$$

Many times in soil engineering we are interested in the stress distribution in or below a dam or a long wall or long footing; in each of these cases the distance along the axis of the structure is great compared with the lateral dimensions. In these situations also all the cross sections are alike, or very similar, so that it is convenient to choose a typical cross section and to study the stresses in it. When all the sections and the stresses acting on them are the same, then no strain or deformation can take place in the direction perpendicular to the cross section, and all movements occur in the plane of the section. This particular type of stress and displacement problem is called *plane strain*. We can develop the special stress-strain relations for it by putting $\epsilon_2 = 0$ in Eqs. (3–18). From Eq. (3–18b) we find that

$$\sigma_2 = \nu(\sigma_1 + \sigma_3) \tag{3-28}$$

and this can be substituted in Eqs. (3–18a) and (3–18c) to give

$$\epsilon_1 = \frac{1}{E} [\sigma_1(1 - \nu^2) - \sigma_3(\nu + \nu^2)] \tag{3–29a}$$

and

$$\epsilon_3 = \frac{1}{E} [\sigma_3(1 - \nu^2) - \sigma_1(\nu + \nu^2)] \tag{3–29b}$$

It is possible to analyze the strains of an element such as that shown in Fig. 3–5 in order to describe the normal and shearing strains in an element oriented at any selected angle to the reference direction. When this is done and the strains are referred to the principal strains in two dimensions, it is found that equations similar to Eqs. (3–9) and (3–10) are obtained:

$$\epsilon = \frac{\epsilon_1 + \epsilon_0}{2} + \frac{\epsilon_1 - \epsilon_3}{2} \cos 2\alpha \tag{3–30}$$

and

$$\frac{\phi}{2} = \frac{\epsilon_1 - \epsilon_3}{2} \sin 2\alpha \tag{3–31}$$

when the angle α is defined and used similarly to the angle α of Eqs. (3–9) and (3–10). The only difference, because of the geometry of the strains and the way ϕ is defined, is the factor 2 in the shearing strain equation, Eq. (3–31).

Now it is possible to substitute the values of ϵ_1 and ϵ_3 from the plane strain case, Eqs. (3–29) in Eq. (3–31) for ϕ, to find the relation between shearing strain and stress. When this is done, we get

$$\frac{\phi}{2} = \frac{1 + \nu}{E} \left[\frac{\sigma_1 - \sigma_3}{2} \sin 2\alpha \right] \tag{3–32}$$

However, the term in brackets in Eq. (3–32) was shown to be equal to the shearing stress τ on the α plane, by Eq. (3–10), so that

$$\phi = \frac{2(1 + \nu)}{E} \tau \tag{3–33}$$

This is actually a general result, and is not necessarily restricted to plane strain, although it was convenient to develop it for that case. Consequently, when a material behaves linearly elastically and all the strains are small, it appears that the normal strains are proportional to the normal stresses only and the shearing strains depend only on the shearing stresses. Consequently, various states of stress and strain can be added together or superimposed to get any desired condition. Usually in Eq. (3–33), a *shearing modulus G* is defined, where

$$G = \frac{E}{2(1 + \nu)} \tag{3–34}$$

so that Eq. (3–33) becomes

$$\phi = \frac{\tau}{G} \tag{3-35}$$

It is seen, from Eqs. (3–18), (3–25), and (3–33), that only two constants, E and ν, are required to describe the relations between stresses and small strains in a linearly elastic isotropic solid. If the material is not isotropic, more constants are required.[2]

When adequate precautions are taken during tests, it is found that sands and gravels will exhibit elastic behavior under applied hydrostatic stresses; but as a result of the stresses at the points of contact between grains and the increasing area of contacts under increasing stresses and strain, the behavior is not linearly proportional to the stress. The behavior of sands under applied shearing stresses is neither linear nor elastic. In clays under both hydrostatic and shearing stresses the behavior is neither linear nor elastic. However, because of the difficulty of analyzing more complicated material behaviors, solutions based on the assumption of linear elastic behavior are commonly used for estimations of stresses in all soil mechanics problems, even though the behavior of no soil may be represented, strictly speaking, as linearly elastic. Some examples of these solutions are given below.

3-6 ELASTIC SOLUTIONS

The equations which must be satisfied in the solution of the stress distribution, strains, and displacements in a linearly elastic solid under stress are the equations of equilibrium, the compatibility equation, and the stress-strain relations of the material. Even in apparently simple problems the solution is not always easy to obtain, and the methods of solving these problems will not be examined here; they are described in textbooks on elasticity. Instead, some solutions of interest in soil engineering will be presented and certain aspects of the results will be discussed. The material in each solution is considered to behave linearly isotropically elastically, and all of the strains are small.

If a concentrated load is applied over a very small area at the surface of a semi-infinite solid (at ground surface in the soil problem), the load can be considered to be a concentrated one of magnitude P and the vertical normal and shearing stress distribution can be obtained from the following equations, which were first obtained by Boussinesq. The equations for σ_x and σ_y are more complicated and will not be given here. With

$$R^2 = x^2 + y^2 + z^2 \quad \text{and} \quad r^2 = x^2 + y^2 \tag{3-36}$$

the Boussinesq equations are

$$\sigma_z = \frac{3P}{2\pi}\frac{z^3}{R^5} \tag{3-37}$$

$$\tau_{rz} = \frac{3P}{2\pi}\frac{rz^2}{R^5} \tag{3-38}$$

In Fig. 3–8a closed contours of equal σ_z stress values have been drawn from Eq. (3–37), and it is seen that these contours have a teardrop or bulb shape. As a consequence, the region within a particular contour surface drawn through points with the same relatively low vertical stress is sometimes referred to as a *pressure bulb*. From the complete set of equations for the stresses for the concentrated load case it is possible to obtain the solution in terms of stresses for a loaded area of any shape on the surface of the semi-infinite solid. In particular, the equations corresponding to Eqs. (3–37) and (3–38) for a line load of intensity p per foot at the surface (plane strain problem) are

$$\sigma_z = \frac{2p}{\pi}\frac{z^3}{R^4} \tag{3-39}$$

$$\tau_{xz} = \frac{2p}{\pi}\frac{xz^2}{R^4} \tag{3-40}$$

when $R^2 = x^2 + z^2$ and x and z are the coordinates in the plane perpendicular to the axis of the load.

From the solution to the problem of stress distribution under a line load, the stresses under a long strip loading of intensity p may also be obtained:

$$\sigma_z = \frac{p}{\pi}[\alpha + \sin \alpha \cos (\alpha + 2\delta)] \tag{3-41}$$

$$\tau_{xz} = \frac{p}{\pi}[\sin \alpha \sin (\alpha + 2\delta)] \tag{3-42a}$$

and

$$\tau_{max} = \frac{p}{\pi}\sin \alpha \tag{3-42b}$$

where the angles α and δ are shown in Fig. 3–8b, in which the contour lines of the two stresses σ_z and τ_{max} are also given. In this figure the region enclosed by the $\sigma_z/p = 0.10$ contour is that usually referred to as the pressure bulb. The contour lines of τ_{max} are all circles, as shown by Eq. (3–42b).

In practical soil engineering problems, loads or stresses are frequently applied to the ground surface over areas which are rectangular in shape, and for these cases it is useful to be able to calculate the vertical normal stress, which is needed for some settlement analyses, in the soil stressed

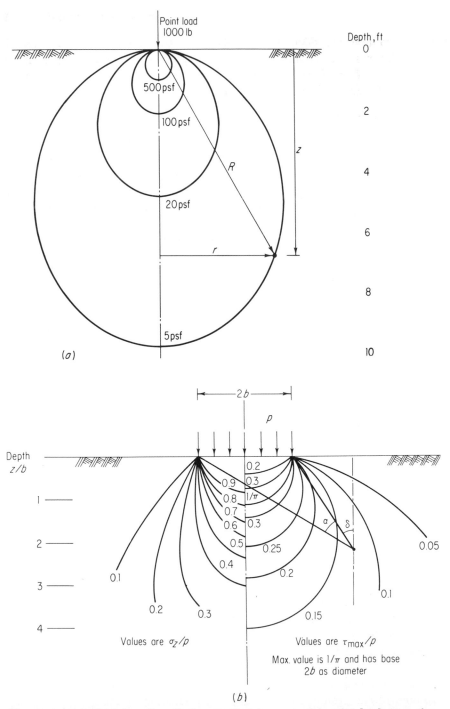

FIG. 3-8 (a) Contours of vertical normal stress σ_z under point load at surface.
(b) Contours of vertical normal stress and maximum shearing stress under uniform
strip load at surface.

by the foundation. Since the equations for the stresses are complicated, it is sufficient to have a means of obtaining the stress at any depth under one corner of the foundation only. For this purpose Fig. 3–9a has been included to give values of the parameter p_0 in the equation

$$\sigma_z = p_0 p \qquad (3\text{--}43)$$

in which p_0 is a function of the width and breadth of the footing when they are expressed in terms of the depth to the point at which the stress σ_z is to be calculated and p is the applied stress on the foundation. It will be recalled that solutions to problems involving small strains in a linearly elastic medium are additive; consequently, the stress anywhere in the soil can be obtained by dividing the foundation up in such a way that the desired point lies below adjacent corners of the subdivided areas. For example, in Fig. 3–9b we want to calculate the stress σ_z at a point P that is 5 ft below the center of the 5- by 10-ft rectangular footing subjected to a uniform pressure of 1,000 psf. If the footing is divided into the four equal areas $2\frac{1}{2}$ by 5 ft shown by the dashed lines, then P lies below their common corner. In each rectangle $m = 0.5$, $n = 1.0$ (they are interchangeable), so that from Fig. 3–9a we find $p_0 = 0.12$. At a depth of

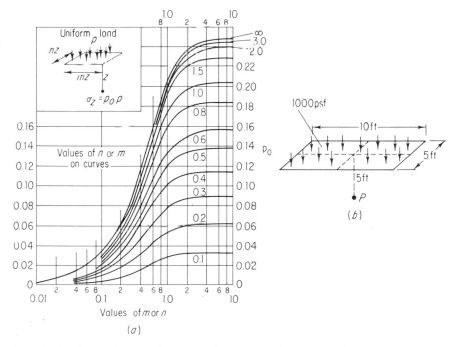

Fig. 3–9 *Vertical normal stress below rectangular uniformly loaded area.* (a) *Chart for calculating vertical normal stress under one corner of a uniformly loaded rectangular area at surface.* (*After Fadum.*) (b) *Example of vertical stress calculation.*

5 ft below the corner of a single $2\frac{1}{2}$- by 5-ft rectangle we therefore have

$$\sigma_z = 0.12 \times 1,000 = 120 \text{ psf}$$

and consequently the stress 5 ft below the center of the original loaded area is given by adding together the stress produced at that depth by the four fictitious smaller rectangles:

$$\text{Total } \sigma_z = 4 \times 120 = 480 \text{ psf}$$

For other points, all the rectangles may have different sizes with varying m and n values, or, in the case of points in the soil outside the loaded area, imaginary rectangular areas must be provided and the final stress must be computed both by adding and subtracting the stress contributions of the different areas. The stress under any shape of loaded area may be found in this way, provided the load may be fairly assumed to be distributed uniformly on the surface.

For problems involving more complicated and more realistic geometries, numerical-solution techniques by computer have been developed. One such method involves dividing the soil region up into triangular pieces which are called *finite elements*. These methods will not be discussed here.

3-7 NONLINEAR BEHAVIOR IN SOILS

In both sand and clay soils the volumetric strains may not be solely dependent upon applied hydrostatic stresses, and volume changes may arise from shearing stresses. Consequently, in soils, a variation of Eq. (3–27) is sometimes used to describe the material behavior in the following form:

$$d_s = \frac{\Delta\sigma_h}{K_s} + \frac{\Delta\tau_{\text{ave}}}{D} \tag{3-44}$$

where d_s is the volumetric strain in the soil structure arising as a result of incremental stress increases $\Delta\sigma_h$ in the hydrostatic stress and $\Delta\tau_{\text{ave}}$ in the shearing stresses applied to the soil structure. The bulk modulus of the soil skeleton is K_s. We shall see the reason for this material description later.

3-8 TOTAL AND EFFECTIVE STRESSES

Let us isolate the section through an element of saturated soil in equilibrium as shown in Fig. 3–10. It will be considered that the two grains in contact represent average grains in a soil and that the stresses indicated in Fig. 3–10 represent average stresses over some region. On some surface in the soil an external stress p, usually called the total or applied stress, acts, and this stress is resisted both by the contact pres-

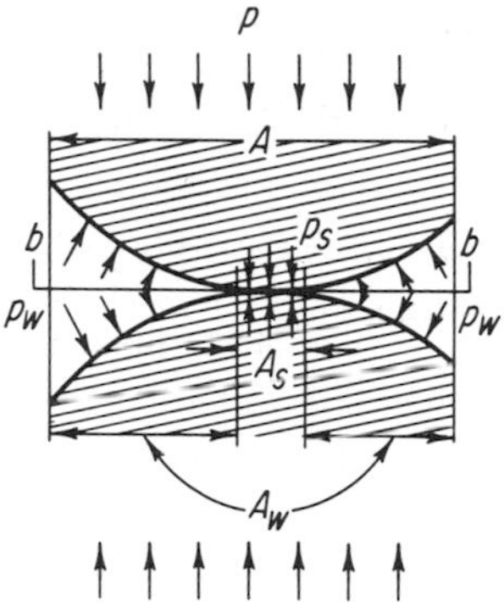

FIG. 3-10   *Soil and water stresses.*

sure p_s between the grains in general and by a hydrostatic pressure p_w in the pore water. All the stresses are measured above atmospheric pressure. Considering the balance of forces acting across the cross section *b-b* passing through the point of contact of the two grains, we can see that

$$pA = p_sA_s + p_wA_w \tag{3-45}$$

The second term on the right-hand side of the equation requires explanation. The stress p_w acts, of course, normal to the surface of the grains, and we find that, if the components of this stress in the direction of p are calculated and summed along the perimeter of either grain in Fig. 3-10, the result is equal to p_w times the horizontal trace A_w of the area of grain surface. Dividing Eq. (3-45) throughout by A and recognizing that $A_w = A - A_s$, we get

$$p = p_s \frac{A_s}{A} + p_w \left(1 - \frac{A_s}{A}\right) \tag{3-46}$$

In this equation the ratio A_s/A is small because of the relatively small area of contact between grains, but, on the other hand, p_s can be a very high stress (in fact, it approaches or equals the yield stress of the solid material of the grains) and therefore the first combination of terms on the right-hand side of Eq. (3-46) is not zero. In the second term p_w is a stress of the same order of magnitude as the applied stress p, and in this case the term in parentheses approaches unity so closely that we can represent Eq. (3-46) in the following way:

$$p = p_s' + p_w \tag{3-47}$$

where

$$p_s' = p_s \frac{A_s}{A} \tag{3-48}$$

In Eq. (3-48) p_s' is a fictitious stress considered to be developed in the soil and is called the *effective stress*. It cannot be measured directly, although the total stress p and the pore-water pressure p_w are both real and measurable pressures in many circumstances. The effective stress is therefore usually calculated from Eq. (3–47) when the total and pore pressures are known.

The stress p_s' represents the part of the total stress which is absorbed or balanced by the soil structure. One of the most important hypotheses in soil mechanics is that the deformation of the soil structure is controlled by this fictitious pressure p_s'. The technique given here and shown in Fig. 3–10 of actually demonstrating by analysis the relationship between the stresses is not strictly correct, but is intended to show that Eq. (3–47) has some physical basis. Alternatively, Eq. (3–47) can be taken as a definition of p_s', the effective pressure in the soil. This relationship in soils was first suggested by Terzaghi and represents a most important step in our understanding of soil behavior. The postulate that the deformational behavior of the soil structure depends upon the effective stresses acting on it is called the *effective-stress hypothesis* (or occasionally Terzaghi's effective-stress principle).

The calculation of effective stress in real circumstances can be illustrated by a number of simple examples. In Fig 3–11 the container on the left is full of soil and water; the water surface coincides with the upper surface of the soil layer. The soil column has length L. A connection has been made from a hole in the bottom of the container (the hole is covered with wire gauze to prevent the soil moving out) to a reservoir of water on the right-hand side.

For the first example the water level in the reservoir on the right is maintained at the elevation a identical with the height of the free-water surface in the soil container. The water is therefore static. We wish to calculate under these circumstances the effective stress in the soil at a depth z below the surface of the left-hand container of Fig. 3–11. The computation proceeds as follows: At a depth z in the soil the total pressure due to the column of water and soil above the level considered is equal to $\gamma_t z$. Note again that all pressures are measured from atmospheric pressure as a base. At the depth z the pressure in the pore water p_w is equal to the unit weight of water times the height z, or $\gamma_w z$, since

Water supply in case(b)

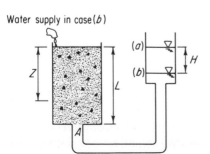

Fig. 3–11   *Illustration   of   effective stresses.*

no flow is occurring. Employing these facts in a rearranged version of Eq. (3–47), we have

$$p'_s = \gamma_t z - \gamma_w z$$
$$= (\gamma_t - \gamma_w)z$$
$$= \gamma_b z \tag{3–49}$$

We see, as might be expected from a commonsense argument, that the effective stress in the soil at any depth is equal to the product of the buoyant unit weight and the depth.

We can make a little more difficult example from the same Fig. 3–11 by lowering the water in the right-hand reservoir to elevation b, a height H below the free surface of water in the left-hand soil container. Water is continually added to the container of soil to maintain the water surface in the left-hand side at its original level. The head difference causes flow through the soil, according to Darcy's law, when the difference is relatively small. As before, the total pressure at the depth z is still due to the column of soil and water above that elevation, and therefore $p = \gamma_t z$ here also. Now the question of calculating the pressure in the pore water is more complicated because of the changing pressure conditions as the water flows through the soil with head dissipation. The pressure head at the surface of the soil is zero (above atmospheric). However, at point A the pressure head on the soil is equal to $(L - H)\gamma_w$. Since the total head varies linearly with distance z through the soil, according to Darcy's law, the pressure head in the water must also vary linearly with distance through the soil, and therefore we conclude that the pressure head at depth z below the soil surface is

$$p_w = \frac{z}{L}(L - H)\gamma_w \tag{3–50}$$

Thus, using Eq. (3–47) again, we can calculate the effective stress to be

$$p'_s = \gamma_t z - \gamma_w \frac{z}{L}(L - H)$$
$$= \gamma_b z + \gamma_w \frac{H}{L} z = \left(\gamma_b + \gamma_w \frac{H}{L}\right) z \tag{3–51}$$

Equation (3–51) shows that the effective stress in the soil at all depths has been increased as a result of the flow of water through the material in proportion to the total head gradient H/L in the soil. Since, in Eq. (3–24) the buoyant unit weight of the soil is a body force per unit volume, it can be seen that $\gamma_w H/L = \gamma_w i$ can also be considered a body force per unit volume as a result of pressure differences in the water flowing through the soil under gradient i. Here the body force due to the water occurs in the same direction as γ_b and therefore adds to the effective stress. Were the flow to occur in the opposite direction, the effective stress would be diminished. Other problems involving calculations of pore pressures and effective stresses can be found in the examples following this chapter.

Since, in general, both pore pressures and effective stresses will be developed in a soil as a result of external applied total stresses, it is possible to substitute Eq. (3–47) into the equations of equilibrium in order to obtain the latter in terms of effective stresses. For this purpose we write Eq. (3–47) in the more general form, for the two-dimensional case

$$\sigma_x = \sigma'_x + p_w \tag{3-52a}$$
$$\sigma_z = \sigma'_z + p_w \tag{3-52b}$$

Since the pore water can sustain only hydrostatic stresses, we observe that the shearing stresses in a soil remain unaffected by the existence of pore-water pressures. On performing the substitution of Eqs. (3–52) in Eqs. (3–4), we obtain

$$\frac{\partial \sigma'_x}{\partial x} + \frac{\partial \tau_{xz}}{\partial z} + \frac{\partial p_w}{\partial x} = 0 \tag{3-53a}$$

$$\frac{\partial \sigma'_z}{\partial z} + \frac{\partial \tau_{xz}}{\partial x} + \gamma_t + \frac{\partial p_w}{\partial z} = 0 \tag{3-53b}$$

Rewriting Eq. (2–1) in the form

$$p_w = \gamma_w h_p = \gamma_w h - \gamma_w h_e \tag{3-54}$$

and taking the derivatives with respect to x and z we get

$$\frac{\partial p_w}{\partial x} = \gamma_w \frac{\partial h}{\partial x} \tag{3-55a}$$

$$\frac{\partial p_w}{\partial z} = \gamma_w \frac{\partial h}{\partial z} - \gamma_w \tag{3-55b}$$

With the x and z coordinates taken horizontally and vertically, respectively, there is no change of h_e (the elevation head) in the x direction, and the rate of change of h_e in the z direction ($+ve$ upward, as in Fig. 3–2b) is unity. Substituting Eqs. (3–55) for the last terms on the left side of Eqs. (3–53), we have

$$\frac{\partial \sigma'_s}{\partial x} + \frac{\partial \tau_{xz}}{\partial z} + \gamma_w \frac{\partial h}{\partial x} = 0 \tag{3-56a}$$

$$\frac{\partial \sigma'_z}{\partial z} + \frac{\partial \tau_{xz}}{\partial z} + \gamma_w \frac{\partial h}{\partial z} + \gamma_b = 0 \tag{3-56b}$$

The terms involving unit weights are then body force per unit volume components due to water pressure components. When seepage is occurring in a soil region, it is, strictly speaking, necessary to solve the flow problem in terms of total heads or seepage gradients everywhere in the soil before the stresses in the system can be analyzed. This is commonly not done. A soil subjected to no stress system other than that due to body forces resulting from water seeping through the pores in, for example, the flow system of Fig. 4–5 would represent a very difficult problem to solve except by numerical techniques using a computer.

3-9 PORE-WATER PRESSURES ARISING FROM APPLIED STRESSES

Soils are composed of two- or three-phase systems in which each of the phases possesses a different volumetric compressibility. Consequently, when an external stress system is applied to a soil element, increases in stress will be registered in the various components of the soil to an extent dependent upon their relative compressibilities.

The pore fluids can develop only hydrostatic pressures. Were the behavior of the soil skeleton ideally linearly elastic, the application of external shearing stresses would have no effect on the volume of soil or on the pore-fluid pressures. In this case only the hydrostatic component of total stresses would develop pore-fluid hydrostatic pressures and a change in the volume of the soil element. However, we have observed in a preceding section that real soils actually exhibit a volume change when subjected to shearing stresses in addition to the volume change undergone by an increase in the hydrostatic pressure. It follows that pore pressures may be developed in the soil pore fluid as a result of the application of external shearing stresses. In practice, this effect is too important to neglect and it is necessary to develop the appropriate relationships in the following manner.

We shall employ Eq. (3–44) to describe the volumetric change behavior of the soil skeleton while recognizing that the normal or hydrostatic stresses used in it must be effective stresses. We shall also assume that a relationship analogous to Eq. (3–27) can be adduced to describe the compressibility of the pore fluid in the form

$$d_v = \frac{\Delta p_w}{K_v} \tag{3-57}$$

where d_v is the volumetric strain in the pore fluid as a result of a pressure change Δp_w. In general, the pore fluid will contain small bubbles of air and the pressure in these bubbles will not be the same as that in the surrounding liquid. However, we have assumed that the bubble gas pressure will be related to the liquid pressure to the extent that a pressure increase in the pore water will result in an equivalent pressure increase in the pore gas. In Eq. (3–57) K_v represents a bulk or compressibility modulus representative of the combined behavior of air bubbles and water.

Considering unit total volume of a three-phase soil system subjected to an increase in total applied stresses, we can see that, in general, there will be a decrease in volume of the soil skeleton dependent upon the effective stresses applied to it by virtue of the effective-stress hypothesis.

In addition, the volume of the pore fluid will undergo a change appropriate to the stress increase Δp_w developed in the pore fluid. Since the soil skeleton and pore fluid are intimately connected, and we assume that no fluid is exuded from the pores as a result of the stress increase, the volumetric change in the pore fluid must be equal to that in the soil skeleton. Thus d_s from Eq. (3–44) must be equal to $n(d_v)$, since the volume of voids in the soil is represented by the porosity per unit total volume. Using Eqs. (3–44) and (3–57) in their appropriate forms gives rise to

$$\frac{n \, \Delta p_w}{K_v} = \frac{\Delta \sigma_h'}{K_s} + \frac{\Delta \tau_{\text{ave}}}{D} \tag{3–58}$$

Here $\Delta \tau_{\text{ave}}$ is the increase in average shearing stresses in the applied-stress system. However, in Eq. (3–58) the incremental increase $\Delta \sigma_h'$ arises from the difference between the increment in the hydrostatic component $\Delta \sigma_h$ of the external applied stress and the pore fluid pressure Δp_w generated by the applied stress according to the effective-stress definition equation (3–47). Consequently, Eq. (3–58) may be rewritten

$$\frac{n \, \Delta p_w}{K_v} = \frac{\Delta \sigma_h - \Delta p_w}{K_s} + \frac{\Delta \tau_{\text{ave}}}{D} \tag{3–59}$$

In Eq. (3–59) we can rearrange the terms to give an expression for the incremental pore-water pressure developed by the increase in external applied stress in the following form:

$$\Delta p_w = \frac{1}{n K_s / K_v + 1} \left(\Delta \sigma_h + \frac{K_s}{D} \Delta \tau_{\text{ave}} \right) \tag{3–60}$$

Now we can examine two limiting conditions of Eq. (3–60): (1) The pore fluid of the soil consists of gas only, i.e., the soil is 0 percent saturated. (2) The pore fluid consists entirely of water and the soil is 100 percent saturated. In the first case, it will be observed that the compressibility of air is much greater than the compressibility of the soil skeleton for all soils. This means that K_v will be very much less than K_s and the term outside the parentheses on the right-hand side of Eq. (3–60) is therefore approximately equal to zero. In this case, and our intuition confirms the conclusion, there will be no pore pressure increment resulting from an increment of applied stress.

In the case of completely saturated soil, it will generally be the case that water is much less compressible than the soil structure, so that K_v is greater than K_s. As a consequence, the term outside the parentheses of Eq. (3–60) will tend toward unity. We see, therefore, that under these circumstances the pore pressure developed by an increment of applied stress will be proportional both to the hydrostatic and shearing components of the applied stress.

The group of terms K_s/D on the right-hand side of Eq. (3–60) is a property of the soil and must be measured experimentally for each soil by measuring the pore pressures developed by various systems of applied stress. If only a hydrostatic stress increase is applied, then the pressure developed in the pore water of a completely saturated soil is equal to the hydrostatic applied stress, and no effective stress increases are generated in the soil.

In general, a load applied to the surface of the ground develops stress states which differ from point to point in the ground as seen in Fig. 3–8, and consequently the pore-water pressures developed vary from place to place. As a consequence of Darcy's law, these variations in pressure cause flow in the pore water. The flow is accompanied by gradually decreasing pore pressures until the applied load is ultimately entirely borne by the increased effective stresses in the soil and the pore pressures are everywhere zero. The increasing effective stresses cause strains and displacements to occur in the soil as a function of time. This process, called *consolidation*, will be discussed in the following chapter. The significance of Eq. (3–60) will be brought out more completely in a later chapter.

PROBLEMS

3-1 At a point in a region of soil under plane strain the stresses are $\sigma_x = +5$ psi, $\sigma_z = +15$ psi, $\tau_{xz} = 3.3$ psi in the direction shown in Fig. 3–3. Determine the magnitudes and direction of the major and minor principal stresses from Eqs. (3–5) to (3–8).

  Answer $\sigma_1 = 16$ psi; $\sigma_3 = 4$ psi; $\theta = 16.7°$ in Fig. 3–3a when AC is the major principal plane.

3-2 Calculate the maximum shearing stress acting on the soil element of Prob. 3–1. *Answer* $\tau_{\max} = 6$ psi.

3-3 Can you tell from the information given in Probs. 3–1 and 3–2 whether the soil is isotropic or not?

  Answer No; because equilibrium equations hold regardless of material properties.

3-4 Draw the Mohr diagram for the element of Prob. 3–1 and check your answers graphically.

3-5 Determine from the Mohr diagram of Prob. 3–4 the orientation of the plane on which the maximum shear stress is acting.

3-6 Points on the Mohr circle of Prob. 3–4 determine stress conditions on different planes through the element of Prob. 3–1. Which point on the Mohr circle represents the plane on which the ratio of the shearing stress to the normal stress is a maximum for the element (called the *maximum obliquity* of stress)?

Answer   It is the point on the circle at which the tangent straight line through the origin intersects the circle.

3-7   What are the shearing and normal stresses and their ratio (shearing/normal) on the plane of maximum obliquity?

3-8   In a deformation test on soil a rectangle such as $OABC$ in Fig. 3–5 is marked out before stresses are applied to the soil. The length OA is 2.0 in., and the length OC is 1.5 in. After the stresses are applied it is found that the corner points O, A, C have been moved by the following amounts:

Point	O	A	C
x displacement u, in.	+0.10	+0.12	+0.13
y displacement w, in.	−0.13	−0.11	−0.14

Calculate the ϵ_x, ϵ_z, and ϕ_{zz} strains on the average over the element.

Answer $\epsilon_x = +0.01$ or $+1$ percent; $\epsilon_z = -0.0067$ or -0.67 percent; $\phi_{zz} = 0.03$ radian or $1.72°$ decrease in angle AOC. AO has extended in length, OC has contracted.

3-9   For a soil with a modulus of elasticity E of 10^4 psi and a Poisson's ratio ν of 0.25 calculate the principal strains in the soil element of Prob. 3–1.

3-10   Calculate the shearing modulus G for the soil of Prob. 3–9 and determine the shearing strain ϕ_{zz} for the conditions of Prob. 3–1.

3-11   Calculate the bulk modulus K for the soil of Prob. 3–9 and compute the volumetric strain in the element of Prob. 3–1.

3-12   On a strip footing of width 4 ft at the ground surface there is a load of 1,000 psf. Determine from Fig. 3–8b the vertical normal stress below the footing center line at depths of 4, 6, and 8 ft below ground surface.   *Answer*   560, 390, 310 psf

3-13   From Eq. (3–39) calculate the vertical normal stresses at the same depths as in Prob. 3–12 below the line of action of a line load with the same intensity of load per unit length (normal to the paper) as the strip footing.   *Answer*   636, 424, 318 psf

3-14   At what ratio of depth to footing width do you conclude from your answers to Probs. 3–12 and 3–13 that the strip footing can be reasonably approximated by a line load?

Answer   Depth to width ratios greater than about 2

3-15   Calculate the vertical normal stress at depths of 4, 6, and 8 ft below the center of a 4-ft-square footing from Fig. 3–9 loaded by 1,000 psf.

Answer   332, 172, 104 psf

3-16   Compare your answer to Prob. 3–15 to values, obtained from Fig. 3–8a or Eq. (3–37), under a point load of the same total magnitude as the footing load of Prob. 3–15. Determine below what depth-to-width ratio the stresses under a square (or circular) footing can reasonably be approximated by those under a point load.

Answer   480, 210, 120 psf. Probably not less than 3.

3-17   If there is a pore pressure of 2.5 psi in the soil element of Prob. 3–1, what are the effective stresses acting? Show this pore pressure on the Mohr diagram of Prob. 3–4.   *Answer*   2.5, 12.5 psi

3-18   Are the shearing stresses in the soil element of Prob. 3–1 changed by the presence of pore pressures?

REFERENCES

1.   S. TIMOSHENKO and G. H. MacCULLOUGH, "Elements of Strength of Materials," 3d ed., D. Van Nostrand Company, Inc., Princeton, N.J., 1949.
2.   J. C. JAEGER, "Elasticity, Fracture, and Flow," 2d ed., Methuen & Co., Ltd., London, 1962.

4 | Seepage and Consolidation

An important aspect of engineering with soil involves the flow of water through the soils associated with various engineering structures. Such a flow takes place, for example, through an earth dam, resulting in leakage from the reservoir, or into a trench cut into soil below the water table. When such a flow occurs, there are two aspects of engineering interest: the quantity of flow and the water pressure at different points in the soil. The first determines the size and layout of pumping and drainage systems, whereas the magnitude and distribution of pressures have important effects on the stability of structures and on the soil strength, as will be seen in Chapter 5.

It was mentioned in Chapter 3 that two different kinds of water flow through soil can be studied: In the first type of flow, termed *steady state*, if we were to measure the pressure or head in the water at a given point, we would find it to remain constant with time. In the second kind of slow, *transient flow*, the pressure in the water at any point in the region in which flow is taking place does not remain constant in time but increases, diminishes, or may fluctuate as a function of time. Steady-state flow conditions prevail when the water flow into any soil element or region equals the flow out of it; if the flows in and out are not equal, some water is stored in or abstracted from the region. The storage can arise for a number of reasons, but the principal one of interest in soil engineering is the compressibility of the soil structure. In a compressible soil, a change in effective stress causes a change in volume and water moves in or out of the material. For a transient-flow situation to develop, therefore, changes in pressures (soil or water) must be occurring in time and the soil structure must be relatively compressible. However, steady-state flow can eventually take place in any type of soil, compressible or

incompressible, if the conditions causing flow remain constant for sufficiently long. We shall take up the study of steady-state flow first and
leave transient flow to a subsequent section.

4-1 STEADY-STATE FLOW

When steady-state flow is occurring, there is a region of soil
through which the flow is taking place and which frequently has external
boundaries of two kinds. On one kind of boundary, the total head or
potential of water is known; this is then termed a *potential boundary*.
The other boundary is delineated by impervious material, such as intact
rock, so that all flow at the boundary must be parallel to it; the name
flow boundary is given to this condition. Other boundaries exist, and they
will be mentioned later. All flow problems of interest to us can be char-
acterized by two potential and two flow boundaries.

We can clarify the discussion by referring to two situations which are
shown in Fig. 4-1. Here we have cross sections through (a) a laboratory
experiment on flow through soil and (b) a sheet-pile wall partially pene-
trating a permeable soil overlying an impermeable bedrock surface. In
the laboratory experiment of Fig. 4-1a the total head of water in the
left-hand reservoir is applied to the soil face AB and the head in the
right-hand reservoir acts across the soil face CD. These two surfaces AB

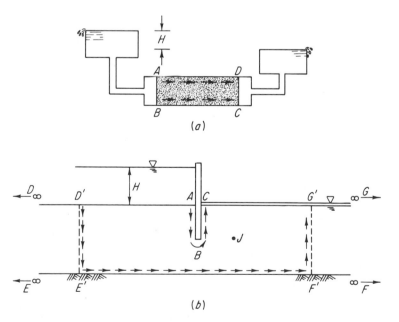

Fig. 4-1 *Boundary conditions for seepage. (a) Simple rectilinear flow. (b) Flow
under a wall; boundary conditions.*

and CD are then the potential boundaries in this configuration. Since the container holding the soil will be impermeable, the water flowing through the soil near the container walls must flow parallel to them, so that AD and BC are the two flow boundaries here.

A more practical situation is shown in Fig. 4–1*b*, where the water flow through the soil is developed by the difference in heads on each side of the wall. Along AD, a constant head acts and AD is therefore a potential boundary in our problem. The head along CG is also constant, and this surface is the second potential boundary. To fulfill its function the sheet-pile wall must be impermeable, and consequently the water flow in the soil near the wall is constrained to be parallel to the wall surface ABC, which becomes one of the flow boundaries. The impermeable surface of the bedrock EF forms the outer flow boundary. If the soil, water, and bedrock conditions extend a long way to the left and right of the wall, then D, E, F, and G effectively lie at infinity. However, in practical circumstances, some change in conditions may occur. If the change is fairly far away from the wall ("far" in this case is of the order of 4 or 5 times the soil layer thickness), we can safely assume that the pressure conditions near the wall and the total quantity of water flowing through the soil will not be changed very much if we ignore the conditions far from the wall.

In practical solutions, therefore, it is usual to assume that the flow occurs within the soil region bounded by the lines $D'E'$ and $F'G'$ taken an arbitrary distance, as mentioned above, from the wall. Because we are assuming that no flow enters the soil region of interest to us from the area outside $D'E'$ and $F'G'$, these are therefore flow boundaries. We can now see by examining Fig. 4–1*b* that $D'E'$ and $F'G'$ are extensions of the bedrock flow boundary, so that, in a practical solution to this problem, the two potential boundaries are AD' and CG' and the two flow boundaries are ABC and $D'E'F'G'$.

We shall now proceed to develop the equations which describe steady-state water flow in soils.

4–2 CONTINUITY EQUATION

For simplicity of representation we shall consider flow in two dimensions only (as in the region of Fig. 4–1*b*). First, we must study the flow of water through a hypothetical soil element $ABCD$ as shown in Fig. 4–2; the element has unit thickness perpendicular to the paper. When three dimensions are taken into account, the equations become a little longer but the following reasoning is in no way changed.

In Fig. 4–2, v, a superficial velocity of water flow (see Chapter 2) through the soil element $ABCD$ can be resolved into the two components v_x and v_z in the positive directions of the coordinate system shown. For

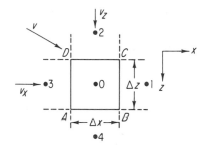

FIG. 4–2   *Components of flow through a finite element.*

the conditions of Fig. 4–2 flow occurs *into* the element through faces AD and DC of lengths Δz and Δx, respectively. If $q_x(AD)$ represents the volume rate of water flowing into the element through the face AD, then

$$q_x(AD) = v_x(AD)\,\Delta z\,1 \tag{4–1}$$

and similar expressions hold for the flow in through face DC and out through faces AB and BC. For steady-state flow to occur, the flow in equals the flow out, so that the following equation can be written:

$$q_x(AD) - q_x(BC) + q_z(DC) - q_z(AB) = 0 \tag{4–2}$$

This is called the *equation of continuity*, and it characterizes steady-state flow through the porous element $ABCD$. Substituting from the various forms of Eq. (4–1), Eq. (4–2) can be written as follows:

$$v_x(AD)\,\Delta z - v_x(BC)\,\Delta z + v_z(DC)\,\Delta x - v_z(AB)\,\Delta x = 0 \tag{4–3}$$

Dividing throughout by $\Delta x\,\Delta z$, we get

$$\frac{v_x(AD) - v_x(BC)}{\Delta x} + \frac{v_z(DC) - v_z(AB)}{\Delta z} = 0 \tag{4–4}$$

Because Δx and Δz are small but still measurable distances, Eq. (4–4) is said to be the finite-difference form of the equation of continuity.

By using calculus, Eq. (4–4) can be written in an alternative form if v_x and v_z are taken to be the components of the superficial fluid velocity at the center of the element. The velocities $v_x(AD)$, . . . can be expressed in the form

$$v_x(AD) = v_x - \frac{\partial v_x}{\partial x}\frac{dx}{2} \qquad v_x(BC) = v_x + \frac{\partial v_x}{\partial x}\frac{dx}{2} \tag{4–5}$$

where the partial derivatives $\partial v_x/\partial x$, . . . represent the spatial rate of change of velocity v_x in the x direction and the differential distances dx, . . . have been employed instead of the finite increments Δx, ... Substituting Eq. (4–5) in Eq. (4–4), we get the following expression for the equation of continuity:

$$\frac{\partial v_x}{\partial x} + \frac{\partial v_z}{\partial z} = 0 \tag{4–6}$$

Before we can proceed further with the equation, we must be able to describe water flow in soil mathematically. It will be seen that, in the derivation of the equation of continuity, no material properties other than the incompressibility of the soil and fluid were assumed and that the soil was taken to be saturated with the fluid. This process of analyzing flow in a soil parallels that of Chapter 3 in which equations were developed to describe stresses and strains in a solid. There equations of equilibrium were first established, and then further equations describing the material behavior were inserted into them.

The necessary mathematical relationship was described and discussed in Chapter 2; it is represented by Darcy's law, which can now be substituted in the continuity equation.

4-3 LAPLACE EQUATION

In order to employ Darcy's law in the form of Eq. (2–5a) we require to know the total heads at the centers of the elements surrounding the element under consideration in Fig. 4–2. These adjacent elements are outlined by dashed lines in that figure, and their centers are numbered as shown, for identification. Then, for example, the constant total head at the center of element 2 becomes h_2.

Using Darcy's law in the form of Eq. (2–5a), in which the general difference Δh is now successively represented by the head differences between adjacent element centers (in Fig. 4–2) $\Delta h = h_3 - h_0$, $\Delta h = h_0 - h_1$, etc., we can obtain expressions describing the velocities $v_x(AD)$, $v_z(BC)$ as follows:

$$v_x(AD) = k_x \frac{h_3 - h_0}{\Delta x} \quad \cdots \quad v_z(AB) = k_z \frac{h_0 - h_4}{\Delta z} \qquad (4\text{--}7)$$

It is assumed that the permeabilities are different in the x and z directions.

Substituting the four expressions of Eqs. (4–7) in Eq. (4–4), we get the following result:

$$k_x \frac{h_1 + h_3 - 2h_0}{(\Delta x)^2} + k_z \frac{h_2 + h_4 - 2h_0}{(\Delta z)^2} = 0 \qquad (4\text{--}8)$$

If Darcy's law in the form of Eq. (2–5b) is substituted in the calculus version of the equation for continuity (4–6), then Eq. (4–8) is obtained in the form

$$k_x \frac{\partial^2 h}{\partial x^2} + k_z \frac{\partial^2 h}{\partial z^2} = 0 \qquad (4\text{--}9)$$

If in Eqs. (4–8) and (4–9) the soil is isotropic, so that $k_x = k_z$, the equations become, respectively,

$$\frac{h_1 + h_3 - 2h_0}{(\Delta x)^2} + \frac{h_2 + h_4 - 2h_0}{(\Delta z)^2} = 0 \qquad (4\text{--}10)$$

and

$$\frac{\partial^2 h}{\partial x^2} + \frac{\partial^2 h}{\partial z^2} = 0 \qquad (4\text{--}11)$$

Equation (4–11) is well known in mathematical physics and is called the *Laplace equation* in two dimensions. Its finite-difference form is given in Eq. (4–10). Since in the finite-difference representation we are free to choose the mesh spacings, we can further simplify Eq. (4–10) by making $\Delta x = \Delta z$ so that

$$h_1 + h_2 + h_3 + h_4 - 4h_0 = 0 \qquad (4\text{--}12)$$

This result will be of value in obtaining solutions to the flow problem and will be used again later. It may be noted from Eq. (4–8) that it could have been obtained even for an anisotropic soil by choosing the mesh spacing to satisfy the condition

$$\frac{k_x \, \Delta z}{\Delta x} = \frac{k_z \, \Delta x}{\Delta z}$$

or

$$\Delta x = \sqrt{\frac{k_x}{k_z}} \, \Delta z \qquad (4\text{--}13)$$

It can be observed from Eq. (4–13) that the anisotropic flow equation in its calculus form, Eq. (4–9), can be reduced to the Laplace equation by transforming the x coordinate of the flow system to an x_t coordinate, where

$$x_t = \sqrt{\frac{k_z}{k_x}} \, x \qquad (4\text{--}14)$$

In most soils problems $k_x > k_z$, so that Eq. (4–14) will usually result in a shortening of the x dimensions of the flow region.

The solution to the Laplace equation in a region is controlled by the boundary conditions to which the region is subjected, and it consists, in its mathematical form, of an equation for h, the total head, in terms of x and z. All other information such as rate of flow and gradients can be obtained from this equation or from its graphical representation. We shall discuss the solution in the next section.

4–4 THE SOLUTION TO THE LAPLACE EQUATION

Flow Nets. Many aspects of solving problems involving steady-state water flow in soil can be clarified by referring to a specific example, and we shall choose the situation of Fig. 4–1b for this purpose. In practice, such a sheet-pile wall is driven as a cofferdam across a river or may encircle a construction area. On one side, the construction side, of the wall, pumps are used to lower the water surface so that dam or bridge foundations can be prepared. The difference in head across the wall then drives the water through the underlying pervious soil. Let us assume that the water table to the right of the wall has been lowered just to ground level, so that the head difference is *H*. We shall assume that a solution to the Laplace equation in the homogeneous isotropic soil region of Fig. 4–1b has been obtained and is shown graphically in Fig. 4–3a by the solid lines. How the solution was obtained will be dealt with later.

The solid lines of Fig. 4–3a are lines of constant total head and are called *equipotentials*. Since the elevation head at any point can be measured with respect to some datum, such as *CG* in Fig. 4–3a, the pressure head in the seeping water can be calculated from Eq. (2–1) for the same point. At a given point in the soil the gradient along any line element will depend on the orientation of the line, with a minimum value equal to zero if the line element is taken tangential to the equipotential through the point, and a maximum if the line element lies perpendicular to the equipotential. No flow therefore occurs in a direction parallel to the equipotential through any point, and the actual direction of flow at a point in an isotropic or transformed soil takes place at right angles to the equipotential line through the point. Lines which intersect all the equipotentials in the

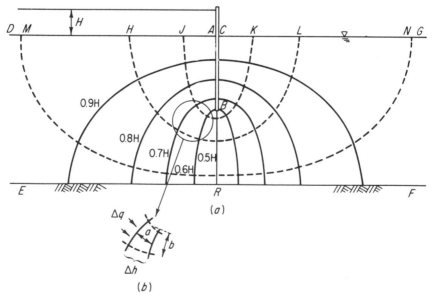

FIG. 4–3   *Solution to flow problem. (a) Equipotentials and flow lines. (b) Single flow rectangle.*

soil region at right angles are called *flow lines*. An infinite number of flow lines can be drawn, but as in the case of the equipotentials, it is convenient to draw only a few representative ones, such as shown by the dashed lines in Fig. 4–3a. From the above argument, it is seen that no flow occurs across a flow line. The graphical solution consisting of flow lines and equipotentials, as shown in Fig. 4–3a, is called a *flow net*.

There are some advantages in selecting particular flow lines to illustrate a solution to the Laplace equation. In Fig. 4–3b we isolate from Fig. 4–3a one of the little areas of soil which is bounded by two arbitrary flow lines and two equipotential lines. Water is therefore flowing between the flow lines in the direction of the small arrows. We shall assume that, since all the angles of intersection of flow lines and equipotential lines are right angles, the figure enclosed is nearly rectangular (we can always obtain a flow element approaching a rectangle by further subdividing the small region of Fig. 4–3b with more flow lines and equipotentials) and Darcy's law in the form of Eq. (2–5a) can be applied to the little region. The head drop between the two adjacent equipotential lines is Δh, so that a quantity of flow Δq takes place between the two flow lines as a result of the head gradient $\Delta h/a$. From Eq. (2–5a) we have

$$\Delta q = k \frac{\Delta h}{a} b \tag{4–15}$$

for unit element thickness perpendicular to the paper. Alternatively, we can write Eq. (4–15) in the form

$$\Delta q = k \, \Delta h \, \frac{b}{a} \tag{4–16}$$

Let us suppose that the two flow lines of Fig. 4–3b intersect one potential boundary AD of Fig. 4–3a in H and J and the other potential boundary CG in L and K. Since no flow can cross the flow lines, the same quantity of flow Δq must flow through any other curvilinear figure formed by the intersection of any two equipotentials and the flow lines H to L and J to K. If, as it happens we have done here, the equipotential lines were drawn representing total heads differing by a constant interval ($0.1H$ in Fig. 4–3a) we can write an equation like Eq. (4–16) for any little figure between HL and JK as follows:

$$\Delta q = k \, \Delta h \, \frac{b}{a} = \text{constant} \tag{4–17}$$

Since k is constant and we have made the head drop Δh constant between any two potentials, it follows that the ratio b/a must be constant for all the little figures such as the figure shown by Fig. 4–3b between the two flow lines. The region between two flow lines is called a *flow path*. If, now, we were to draw another flow line MN, say, it would be convenient to choose it in such a position that the quantity of flow Δq between MN and HL was equal to that between HL and JK. By Eq. (4–17) it follows that the b/a ratio for all the new flow figures formed would be constant and equal to the previous value.

By continuing this process, the entire soil region could be filled in with flow lines so located that the flow between any pair of them was constant and equal to Δq. Only between the last such line, say MN in Fig. 4–3a, and the boundary EF would the b/a ratio of the little flow figures be different, in general, from that in the rest of the diagram. Even there, because of the constant quantity of flow between any pair of flow lines, the ratio would be constant along the whole flow path.

We are now in a position to calculate the total quantity of water flow through the soil region from the diagram, Fig. 4–3a. It should be pointed out that we have assumed so far that the equipotential lines were obtained from a solution of the Laplace equation for this problem. If the solution were in mathematical form, the flow quantity could, of course, also be expressed mathematically. However, for most practical problems of flow through soils, the mathematical solution is not obtained, so that the accompanying technique is most useful. The total flow q can be obtained by summing up the flows occurring in all the flow paths as follows:

$$q = \sum \Delta q = \sum k \, \Delta h \, \frac{b}{a} = k \, \Delta h \, n_f \frac{b}{a} \qquad (4\text{–}18)$$

where n_f is the number of flow paths. Because partial flow paths occur, such as that between MN and the boundary EF in Fig. 4–3a, n_f will not in general be a whole number. We can also represent the incremental head drop Δh as a result of dividing the total head drop H (in general $H_2 - H_1$, where H_2 and H_1 are the two boundary total head values) by the number n_d of potential head drops we have chosen to include in the figure. Thus, Eq. (4–18) can be rewritten

$$q = kH \frac{n_f}{n_d} \frac{b}{a} \qquad (4\text{–}19)$$

Thus, by sketching in the solution to the Laplace equation in the region of interest, we can obtain a good estimate of the total seepage flow quantity with a minimum of further calculations.

In Eq. (4–19), k and H are given by the conditions of the problem, but n_f/n_d and b/a, or alternatively, their product, are dimensionless numbers appearing as a direct result of the solution. It can be seen that, if the solution is obtained in the form of equipotential lines, flow lines can be drawn in for any selected value of the ratio b/a; however, the value chosen determines the number n_f of flow paths obtained (since the boundaries and the equipotentials are fixed), so that the product $(n_f/n_d)(b/a)$ is a constant for a given geometrical shape of the region in which flow is occurring. It is therefore sometimes called a *shape factor*.

We have seen that the ratio b/a of the dimensions of the flow figure of Fig. 4–3b can be varied arbitrarily, and we might choose a value for this ratio which facilitates the development of a solution. An obvious ratio is unity, so that the flow element is square or, more precisely, a curvilinear square. We have demonstrated that, if the head drop between equipotentials is constant, all flow elements along a given flow path must have the same b/a ratio—unity, in the present case—and that it is possible to select other flow lines to give the same quantity of flow along all flow paths (except that next a boundary). Consequently, the following conditions must be met by our solution in its simplified form in the isotropic (or transformed) region.

1. All flow elements are square.
2. All flow lines must meet equipotential lines at right angles (boundary flow lines may not meet boundary potentials at right angles).
3. No flow may occur across a flow line.

The technique of solving the problem is then to sketch flow and equipotential lines in the soil region while trying to meet these conditions as well as possible. Figure 4–4 illustrates different stages in the sketch solution of the problem of Fig. 4–3a. One begins by drawing a small section

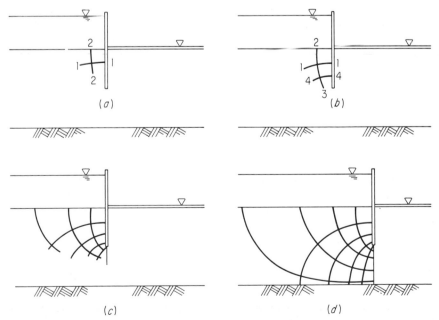

FIG. 4–4   *Successive stages in drawing a flow net.*

of a flow or equipotential line at some convenient point, in this case, say, potential line 1-1 next to the sheet-pile wall, as in Fig. 4–4a. No attention is paid to the value of the potential, since it can be calculated after the flow net is completely formed and complies with the boundary condition. Then a section of the equipotential or flow line, say 2-2 in Fig. 4–4a, is sketched in to form an element as nearly square as possible. The next stage is shown in Fig. 4–4b, where the line 2-2, for example, is extended slightly down the wall to 3 and a piece of the next equipotential 4-4 is inserted to form a second square 1-1, 4-4. The flow and equipotential lines are extended and more lines are added as in Fig. 4–4c to fill the soil region to the final solution in Fig. 4–4d.[1] The last flow line must be compatible, in this sample problem, with the lower flow boundary. Only one-half of the symmetrical solution is shown in Fig. 4–4.

Provided the three listed conditions are met to a reasonable extent and the solution conforms to the boundary conditions, flow quantities calculated from the sketched net with the use of Eq. (4–19) can be obtained to accuracies of 10 to 20 percent in many problem configurations, although values of head or pressure at a given point may be in error to a much greater extent.

Frequently, flow occurs in regions consisting of layers or lenses of soil of different permeabilities. When it does, we must have a mathematical expression to describe the flow condition at the boundaries between different soils. The correct equation will be given here without proof; it is obtained by considering the continuity of the normal component of flow across the boundary. In Fig. 4–5a a boundary between two regions of permeabilities k_1 and k_2 is shown, together with a portion of a flow line intersecting the boundary at angles α_1 to the normal in the k_1 material, and α_2 to the normal in the k_2 material. The angles α_1 and α_2 are related by the equation

$$\frac{\tan \alpha_1}{\tan \alpha_2} = \frac{k_1}{k_2} \tag{4–20}$$

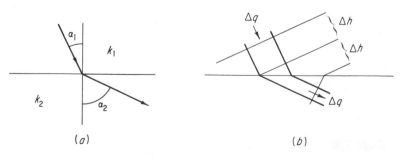

(a)   (b)

FIG. 4–5   *Flow in soils of different permeabilities. (a) Flow line at interface. (b) Flow path at interface.*

Since equipotential lines lie at right angles to the flow lines in isotropic soil, another relation like Eq. (4–20) must govern the entry and exit angles of equipotential lines at the boundary.

In Fig. 4–5b a flow path is shown crossing the boundary between soils of permeabilities k_1 and k_2. Because the rate of flow in the path must be the same in both regions, we have the requirement

$$\Delta q_1 = k_1 \frac{\Delta h_1}{a_1} b_1 = k_2 \frac{\Delta h_2}{a_2} b_2 \tag{4–21}$$

where the subscripts refer to the flow elements in the two regions. If, as shown in Fig. 4–5b, $\Delta h_1 = \Delta h_2$ and the flow figures are square in region 1, then the elements cannot be square in region 2, and their proportions b_2/a_2 are given by Eq. (4–21). It is possible to make them square in both regions, provided that Δh_2 and Δh_1 in this case are related by Eq. (4–21). If the latter alternative is selected, additional equipotentials are required in one of the regions. It will be seen that a problem involving regions of different permeabilities becomes extremely difficult to solve by the sketching method.

The Laplace equation can be solved exactly mathematically only in relatively simple cases; complicated problems require mathematical skill and usually involve an inordinate amount of labor.[2] When, in a practical problem, a more exact solution than is possible or feasible by sketching is required, either numerical techniques based on Eq. (4–12) or analog methods may be employed. Because of the uncertainty associated with the determination of soil properties, general use is made of these latter two methods by soil engineers, and techniques of obtaining mathematical solutions will not be described here.

It must be emphasized that the omission of the analytical techniques at this stage is occasioned by the limited scope of this text and does not imply that mathematical solutions, as such, have no value in practice. The lack of use of mathematical methods in soil engineering arises in part because of the extremely difficult nature of the conditions involved in real soil problems and also in part because many soil problems are unique, with boundary conditions which cannot be changed by the engineers. Thus a solution is usually required for one set of unalterable circumstances only. Exact mathematical techniques are generally inefficient in such cases, and they are best employed in problems where the engineer or designer has at his disposal a number of parameters which he can vary to obtain a desirable or economical design. Then in the algebraic form of the general mathematical solution, the engineer can substitute different values of the variables to see the effect on the design of changing them. Here, then a general solution is obtained once by the expenditure of some effort; in it changes of variables can be made with relative ease to study particular cases. For this type of problem, in general, numerical or analog methods are inefficient, since a new solution must be constructed for each variation of the general problem.

As an example, a mathematical solution is available for the case of Fig. 4–3a generalized to the extent of taking into account any depth of penetration of the sheet piling when the soil is homogeneous. (No layers of differing permeabilities are present.) If this situation should occur at some site in practice, the designer of the sheet-pile wall will probably have information available on the costs of piling driven to various depths and the cost to buy, or rent, and operate pumping equipment of different capacities. The problem facing him is that the deeper he drives the sheet piling, the more costly it becomes; on the other hand, he will spend less on pumping water on the construction side of the wall. The engineer can, in this case, usefully employ the known mathematical solution to construct a graph of quantity of seepage (and thus pumping cost) versus depth of penetration of the piling. On the same chart he can plot the piling costs for different penetrations and thus find out at what penetration his total costs are minimized. In this discussion, consideration of the stability of the piling under the loads acting on it has been left out for clarity.

On the other hand, if, for various reasons, the piling can only be driven to a depth determined by other considerations and the soil contains layers of material of different properties, the designer would still like a good estimate of seepage quantity, and this can best be obtained by carrying out a numerical or analog solution to his unique problem. These latter methods will now be discussed.[3]

Numerical Method.   Here the technique consists in covering the flow region of the problem, such as that shown in Fig. 4–6, with a rectangular network of lines at some convenient spacing, depending on the accuracy required. The Δx and Δz spacings are equal if the soil is isotropic or are related by Eq. (4–13) if it is not. By means of a rough flow-net sketch or guesswork, initial values of the head are written down at each of the network points. For convenience in numbering, a total head difference of 100 or 1,000 is usually selected; the actual value can be used later. When this is done for the flow region of Fig. 4–6, it is seen that, as before, a line of symmetry exists and that we therefore have no need to solve the problem in the two regions on each side of it. It will be sufficient to obtain a half solution which can be duplicated. With a total head difference of 100 the line of symmetry must have a potential of value 50. We therefore choose to solve the problem in the region on the right of the axis of symmetry.

Of the variations of numerical analysis available for solution we shall describe the so-called iteration method. We have shown previously that Eq. (4–12) is a finite-difference form of the Laplace equation; this means

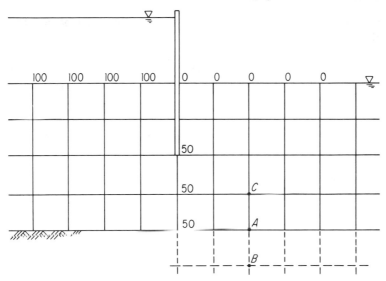

FIG. 4-6   *Numerical analysis of flow.*

that, if the numerical values of head at each point in the network we have drawn satisfy Eq. (4-12), the values represent the solution to the Laplace equation in the region. For this method, Eq. (4-12) is rewritten in the form

$$h_0 = \tfrac{1}{4}(h_1 + h_2 + h_3 + h_4) \tag{4-22}$$

The technique consists in applying Eq. (4-22) to each point of the grid in turn, using the selected initial values to determine successively a new value for the head at the point. The convention of Fig. 4-2 is used in numbering the points around each node. When this procedure has been carried out at all points, we return to the first point and reiterate the operation until total head values which are not altered substantially by further trial computations have been achieved. These will not be exactly correct results, but they can be made as close as may be desired to final values by increasing the number of computations. Usually calculation of the head at each point until the changes are within 1 to 5, based on a head difference of 100, is sufficient.

In order to carry out the iterative process at points such as A in Fig. 4-6 lying on an impervious boundary, we recall that the flow normal to an impervious boundary is zero, so that there can be no flow along the line CA of Fig. 4-6. To achieve this we assume a fictitious point B on the grid on the other side of the boundary and attribute to B the same (fictitious) head as that which exists at point C. When Eq. (4-22) is used here in the determination of the head at A, the term h_2 (the head at C)

will be employed twice. Other numerical methods have been developed in forms suitable for solution by computers.

Electrical Analog.   In the analysis of two-dimensional electrical flow through a conducting medium it is also found that the Laplace equation describes the flow process when the voltage (potential) at any point in the medium replaces the total head h of Eq. (4–11). In these circumstances electrical flow is analogous to water flow through a porous medium, and this fact can be employed to give a convenient solution to a water-flow problem.

In this method an electrically conducting paper is employed to represent the soil region through which flow is occurring. Since the electrical and seepage situations are exactly analogous, the solution involves cutting a piece of the conducting paper to form a scale model of the flow region. Along the potential boundaries an extra strip of paper is left, and it is painted with a highly conducting paint which will give a uniform potential along the boundary. A cut edge of the paper forms a flow boundary, since, of course, electrical flow cannot take place normal to such an edge. When the flow region extends effectively to infinity in practice, the electrical analog is limited by cutting the paper along lines such as $D'E'$ and $F'G'$ in Fig. 4–1b, as is done in other methods of solution.

A solution is obtained by applying suitable electrical potentials (a difference of 5 to 10 volts is usually convenient) to the potential boundaries and using a Wheatstone bridge, potentiometer, or vacuum-tube voltmeter to determine the voltages in the region through which flow is occurring. If a certain potential is selected, the probing terminal can be made to follow the path on the paper along which the potential is constant; the path is then an equipotential line. If scale relationships are derived between the electrical model and the hydraulic prototype, the current flow through the model can be measured and converted to the rate of seepage.

4–5 SEEPAGE FORCES

If the soil container of Fig. 4–1a is turned so that flow is taking place vertically upward through it as in Fig. 4–7, it is instructive to examine the balance of forces on the soil. On the base and top of the soil, respectively, the static pressures in the water are $h_1\gamma_w$ and $h_2\gamma_w$.

The force F exerted on the soil mass by the water is then equal to the difference in the two pressures acting over the cross-sectional area of the soil:

$$F = A\gamma_w(h_1 - h_2) \tag{4–23}$$

Since it can be seen from Fig. 4–7 that $h_1 = H + h_2 + L$, Eq. (4–23)

can be rewritten

$$F = A\gamma_w(H + L) \tag{4-24}$$

In the static case, when no flow is occurring, $H = 0$ and the force is then F_0, where $F_0 = A\gamma_w L$. Therefore the additional force acting on the soil as a result of seepage is given by the difference $F - F_0$ as follows:

$$F - F_0 = A\gamma_w(H + L) - A\gamma_w L = A\gamma_w H \tag{4-25}$$

It is convenient to divide this by the volume of the soil AL to get a seepage force per unit of volume J:

$$J = \frac{F - F_0}{AL} = \gamma_w \frac{H}{L} \tag{4-26}$$

Thus, as a result of water flow through the soil, there arises a seepage body force per unit of soil volume equal to the unit weight of water times the seepage gradient, as shown also by Eq. (3-51). From the solution to a flow problem obtained by any of the methods described previously, the seepage gradient at any point in the region can be calculated by measuring the scale distance on the flow net between any two points of interest at which the total heads can be determined. The ratio of the head difference between the points to the prototype distance between them gives the gradient. Measurement of the gradients is important for the following reasons.

The total head gradient can be increased until the seepage force per unit of volume is just equal to the buoyant unit weight of the soil. At this gradient the soil is just on the point of being moved vertically out of its container; at the same time the effective stress between the soil grains has become zero. If the soil is cohesionless, it will therefore have no strength (as we shall see in Chapter 5) at this gradient, which is referred to as the *critical gradient* i_c. We have

$$\gamma_w \left(\frac{H}{L}\right)_c = \gamma_w i_c = \gamma_b$$

or

$$i_c = \frac{\gamma_b}{\gamma_w} = \frac{G_s - 1}{e + 1} \tag{4-27}$$

FIG. 4-7 *Flow vertically upward through soil.*

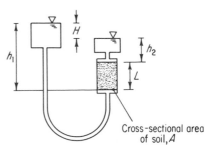

Cross-sectional area of soil, A

Alternatively, the same process can be viewed in the following way. If the reservoir to the left of the figure is raised, a level can be reached at which the force due to the water pressure difference $A\gamma_w(h_1 - h_2)$ just balances the weight of the soil. Let us assume that this height has been reached in Fig. 4–7 and write the equation for the balance of forces across the base of the soil

$$A\gamma_w(h_1 - h_2)_c - AL\gamma_t = 0 \tag{4–28}$$

Rearranging Eq. (4–28), we find that

$$\gamma_w\left(\frac{H}{L}\right)_c = \gamma_t - \gamma_w = \gamma_b$$

as before. When the soil becomes unstable in this way, we say that it has become "quick," or that a "quicksand condition" has developed.

For many soils, the specific gravity is in the range 2.6 to 2.8, and void ratios of 0.6 to 0.8 quite commonly occur. Substituting these values in Eq. (4–27), we can see that the critical upward vertical gradient is close to unity in many cases. Critical upward gradients of this magnitude are very undesirable in connection with any engineering structure. Such conditions can develop, for example, on the right-hand side of the sheet-pile wall in Fig. 4–3a, where water flow takes place vertically upward. In this case a critical value of the head H on the left-hand side of the wall could be found such that the gradient next to the wall on the other side would be sufficient to cause the soil there to become quick. In this state, the soil in the quick region would have lost the ability to support the wall and, as the soil moved out, a stage at which the wall would fail would be reached. In actual fact, since the loss of strength in a cohesionless material would be proportional to the seepage gradient through the soil at the right-hand surface, failure of the wall would take place *before* the head on the left-hand side had reached the critical value for a true quicksand condition to develop. These conditions can arise in connection with any soil and water situation in which seepage with an upward component occurs through some region. Either the water head or flow conditions must be changed or the preventive measure of building an *inverted filter* must be taken to ensure against the loss of soil on the right-hand side of structures such as shown in Fig. 4–3a. Such precautions are described in Chapter 10.

Equation (4–26) represents the special case in which the direction of seepage is vertical. However, as a result of the pressure differences in the water, a body force in a direction normal to the equipotential lines is exerted at every point in the soil mass in which pore pressure differences occur. This force is equal to the unit weight of water times the gradient normal to the equipotential and is therefore directed tangentially to the flow line through the point in an isotropic soil.

Unconfined Flow. We have chosen to illustrate our discussion of seepage through soil with the example shown in Figs. 4–1b and 4–3. In that situation flow was constrained by real boundaries (impervious rock, sheet-pile wall) and the flow is said to be *confined.* However, in some cases of flow of interest to soil engineers, flow in certain regions may not be confined, and a different boundary condition occurs. This type of problem is illustrated by flow through the homogeneous earth dam of Fig. 4–8a and by flow to the well of Fig. 4–8b.

In the case of the earth dam, it is desirable to prevent any seepage through the downstream face, and to this end, drainage layers of coarser soil are placed within the dam as shown by *CD* in Fig. 4–8a. With flow into such a drain, there is a possibility that some of the finer soil grains in the dam may be washed into the drain, with a consequent loss of strength to the dam. To guard against this, the drain is carefully designed and consists of layers of successively finer soil from its base *CD* up, with the requirement that no soil from one layer may pass through the soil of the next coarser layer. The layer of finest material adjoins the soil of the dam itself.

In Fig. 4–8a and b and in similar conditions, the line *AB* will be recognized to be a flow boundary because next to it, in the flow region, flow must be parallel to it, but it must also be a line (or surface) on which

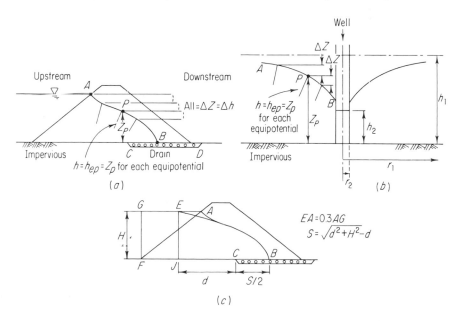

FIG. 4–8 *Unconfined flow. (a) Flow through earth dam. (b) Flow to a well. (c) Construction for free-surface line.*

the atmospheric pressure acts. It is called a free surface or phreatic surface. Thus, since we are using atmospheric pressure as a zero datum, the total head at any point on the surface AB must be equal to the elevation of the point measured from the elevation datum. If CD is such a datum in Fig. 4–8, the value of the equipotential passing through point P, for example, must equal z, the height of P above the plane CD. Since we usually choose to draw equipotentials at equal intervals of head, the points of intersection of these equipotentials with the free surface must be separated vertically by equal amounts which also equal the total head difference between each pair of equipotentials, as shown in Fig. 4–8.

For a practical case, it is not easy to find the position of the free surface, since it is controlled only by the above requirement that it be a flow line on which only atmospheric pressure acts, and the usual procedure is to establish it by a trial-and-error method based, for a homogeneous soil, on the approximate construction, shown in Fig. 4–8c, for an earth dam or embankment on an impervious base. This technique is derived from the knowledge that the free surface closely approximates a parabola over much of its length.

For the earth dam of Fig. 4–8a or c it can be shown that the rate of seepage q is given approximately by the equation

$$q = kS \tag{4–29}$$

where S is shown on Fig. 4–8c.

The flow quantity q into the well of Fig. 4–8b can also be given

$$q = \frac{k\pi(h_1{}^2 - h_2{}^2)}{\log_e (r_1/r_2)} \tag{4–30}$$

where h_1 is the height of the water surface above the datum plane at a distance r_1 from the well, h_2 is the elevation of water in the well, and r_2 is the radius of the well.

4–6 CONSOLIDATION

The left-hand side of Eq. (4–4) represents the net rate of flow in two dimensions into or out of a soil element of sides Δx and Δz; in the preceding sections on steady-state flow the right-hand side of the equation was made zero. Now we must consider the case of transient flow arising when the right-hand side is not zero. The right-hand side will now represent the rate of change of volume of water stored in the element. We shall restrict our discussion to a saturated soil, and in this case the removal (or addition) of a quantity of water Δq from the element will result in a change in volume of the element and therefore necessarily of the soil structure in the element. The process of flow involving volume change in the soil as a function of time is called consolidation.

4-7 CONSOLIDATION EQUATION

The change in volume of the soil structure is equal to the volumetric strain increment Δd_s times the volume, giving $-\Delta d_s\,(\Delta x\,\Delta y\,\Delta z)$, and this occurs in time Δt, so that the rate of change of volume R is given by the equation

$$R = -\frac{\Delta d_s}{\Delta t}\,(\Delta x\,\Delta y\,\Delta z) \tag{4-31}$$

When contraction in volume is occurring, it is usual to consider that the soil skeleton behaves linearly elastically, since, for the most part, only small strains of any element are involved in consolidation. This also simplifies the mathematics considerably and enables us to obtain convenient solutions to a wide range of problems. In this case we may use Eq. (3–27) in terms of the hydrostatic effective-stress increment $\Delta(\sigma'_x + \sigma'_y + \sigma'_z)/3$, and we can substitute effective stresses for Δd_s in order to express the rate of change of volume in the following way:

$$R = -\frac{\Delta(\sigma'_x + \sigma'_y + \sigma'_z)}{3K\,\Delta t}\,(\Delta x\,\Delta y\,\Delta z) \tag{4-32}$$

However, by Eq. (3–52), effective stresses can be expressed in terms of total stresses and pore-water pressure and Eq. (4–32) becomes

$$R = -\frac{\Delta(\sigma_x + \sigma_y + \sigma_z) - 3\Delta p_w}{3K\,\Delta t}\,(\Delta x\,\Delta y\,\Delta z) \tag{4-33}$$

In many problems of practical interest it is found that the time during which the external load is applied (the construction period) is short compared to the time of transient pore-water pressures and settlements. To all intents and purposes, therefore, in such situations the total stresses can be considered to be constant in time, so that $\Delta(\sigma_x + \sigma_y + \sigma_z) = 0$ and Eq. (4–33) reduces to

$$R = +\frac{\Delta p_w}{K\,\Delta t}\,(\Delta x\,\Delta y\,\Delta z) \tag{4-34}$$

Since in the present context we are discussing the application of stresses to the soil and the resulting change of stresses and pressures in the soil structure and pore water, it is appropriate to recast Eq. (4–10) in the form of pressure rather than total head. In addition, because we know that no flow will occur if the pressure is everywhere hydrostatic, we can divide the total pressure in the pore water into hydrostatic, or steady state, p_h and hydrodynamic or excess over hydrostatic, u, components:

$$p_w = p_h + u \tag{4-35}$$

It is the hydrodynamic pressure which causes flow in the pore water in the present circumstances, and the resulting flow will continue until all pressures are hydrostatic or have reached appropriate steady-state values. Equation (2–1) gives the relation between total head, elevation head, and pressure head; with Eq. (4–35) above, it gives

$$h = h_e + h_p \tag{2-1}$$

$$= h_e + \frac{1}{\gamma_w} (p_h + u) \tag{4-36}$$

In the left-hand side of Eq. (4–10) or (4–11), the total head h appears as a second difference or derivative so that, in the present case, the second difference in the z direction becomes, for example,

$$\frac{1}{(\Delta z)^2} (h_2 + h_4 - 2h_0) = \frac{1}{(\Delta z)^2 \gamma_w} (u_2 + u_4 - 2h_0) \tag{4-37}$$

and the second *derivative* takes the form

$$\frac{\partial^2 h}{\partial z^2} = \frac{1}{\gamma_w} \frac{\partial^2 u}{\partial z^2} \tag{4-38}$$

since only the hydrodynamic component of head can be differentiated twice.

With this information the continuity equations (4–8) and (4–9) for the two-dimensional case can be rewritten to include storage:

$$\frac{k_x}{(\Delta x)^2 \gamma_w} (u_1 + u_3 - 2u_0) + \frac{k_z}{(\Delta z)^2 \gamma_w} (u_2 + u_4 - 2u_0)$$

$$= \frac{\Delta u}{K \, \Delta t} \tag{4-39}$$

In the *differential* form the equation becomes

$$\frac{k_x}{\gamma_w} \frac{\partial^2 u}{\partial x^2} + \frac{k_z}{\gamma_w} \frac{\partial^2 u}{\partial z^2} = \frac{1}{K} \frac{\partial u}{\partial t} \tag{4-40}$$

If the material is isotropic with respect to permeability, $k_x = k_z = k$, the material parameters k, γ_w, and K are usually collected to give a *coefficient of consolidation* c, where

$$c = \frac{Kk}{\gamma_w} \tag{4-41}$$

Using Eq. (4–41), Eqs. (4–39) and (4–40) become respectively

$$\frac{c}{(\Delta z)^2} (u_1 + u_2 + u_3 + u_4 - 4u_0) = \frac{\Delta u}{\Delta t} \tag{4-42}$$

and

$$c \left(\frac{\partial^2 u}{\partial x^2} + \frac{\partial^2 u}{\partial z^2} \right) = \frac{\partial u}{\partial t} \tag{4-43}$$

These equations therefore describe the change in hydrodynamic or excess pore-water pressure with time due to the flow of water out of soil under an applied external stress when drainage is present at the boundaries of the soil mass. As the water flows out, the pore-water pressure decreases and the external stress is increasingly transferred to the soil structure as effective stress. The soil consequently undergoes a decrease (usually) in volume, and this is referred to as consolidation.

In its one-dimensional form the consolidation equation was first derived by Terzaghi for saturated soil in the early 1920s and is therefore sometimes called the Terzaghi consolidation equation. The equation is identical with that obtained to describe transient *heat* flow if pore-water pressure and coefficient of consolidation in the soil-consolidation problem are replaced by temperature and thermal diffusivity in the heat-transfer situation. Other transient flow-storage situations can be stated in terms of analogous equations, so that the consolidation equation is a member of a family of so-called diffusion equations. Equations (4–42) and (4–43) can be written in their one-dimensional forms and, in fact, the three- and two-dimensional equations are relatively rarely employed in practice because of the mathematical difficulties involved in obtaining solutions to them.

Some features of the consolidation equation, its solutions, and their employment in various circumstances can best be understood by examining a particular problem of one-dimensional consolidation. We shall study what happens when a circular disk of saturated compressible soil confined radially by an unyielding metal ring is subjected to a stress applied in the axial direction. This is the method used in practice to determine both the compressional properties of a soil and its coefficient of consolidation, and we refer to it as a consolidation test.

4-8 CONSOLIDATION TEST

A schematic diagram of a consolidation cell is shown in Fig. 4–9. The circular disk of soil is held by the metal ring and is in contact with two porous disks top and bottom; the permeability of the disks is much greater than that of the soil, and they are essentially incompressible. First, we shall examine the conditions a very short time after the application of the total vertical external stress $\Delta\sigma_z$ (a principal stress) to the top porous disk. The time interval is chosen small so that, in effect, no pore-water movement out of the soil can have taken place up to the instant we are studying. In Chapter 3 we have seen that, in a saturated soil, the pore water may be considered much less compressible than the soil and in fact may be taken to be incompressible. Thus the volumetric strain of the soil sample of Fig. 4–9 upon application of $\Delta\sigma_z$ is zero. Referring to Eq. (3–25), it will be seen that, for zero volumetric strain,

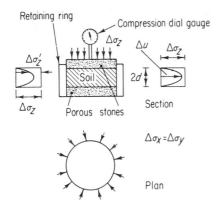

FIG. 4 – 9　*One-dimensional consolidation test.*

Poisson's ratio ν of the mass of saturated soil is equal to $\frac{1}{2}$. In addition, because of the axial symmetry of the sample, the two lateral total (principal) stresses $\Delta\sigma_x$ and $\Delta\sigma_y$ must be equal, and by this symmetry and the rigidity of the confining ring the lateral strains ϵ_x and ϵ_y are both equal to zero. Putting these facts in Eq. (3–18) shows that

$$\Delta\sigma_x = \nu(\Delta\sigma_z + \Delta\sigma_y) \qquad (4\text{--}44)$$

but

$$\Delta\sigma_x = \Delta\sigma_y \qquad \text{and} \qquad \nu = \tfrac{1}{2}$$

therefore

$$\Delta\sigma_x = \tfrac{1}{2}(\Delta\sigma_z + \Delta\sigma_x) \quad , \text{ or } \qquad \Delta\sigma_x = \Delta\sigma_z$$

and therefore

$$\Delta\sigma_y = \Delta\sigma_x = \Delta\sigma_z$$

Thus the saturated soil under these circumstances is subjected to a hydrostatic total stress equal to $\Delta\sigma_z$.

Again because of the volumetric rigidity of the pore water in comparison with that of the soil structure, we saw from Eq. (3–60) that the pore-water pressure generated by an hydrostatic stress applied to a saturated soil was equal to the applied pressure. Consequently, the application of the vertical stress $\Delta\sigma_z$ to the saturated soil in the idealized conditions of Fig. 4–9 generates no effective stresses, and the increase in the pore-water pressure is equal to the applied vertical stress.

We have now determined an initial condition for the solution of the consolidation equation within the boundaries of the soil sample. In practice, as a consequence both of inexact trimming and possible incomplete saturation of the sample, an initial compression of the soil is always observed. We shall assume for the present that an initial value of pore pressure Δu closely equal to the applied stress $\Delta\sigma_z$ is developed.

Let us at this stage consider what will be the final conditions in the sample. When the pore pressure generated by the applied stress has dissipated by drainage of some pore water out through the porous stones, a contraction in volume of the sample equal to the volume of water drained out will have occurred. This contraction, because of the presence of the confining ring, will take place entirely in the vertical direction. All stresses in the soil will now be effective stresses. Again because of axial symmetry $\Delta\sigma_x' = \Delta\sigma_y'$ and because of the ring, $\epsilon_x = \epsilon_y = 0$, so that, from Eqs. (3–18b) and (3–18c)

$$\Delta\sigma_x' = \Delta\sigma_y' = \frac{\nu_s}{1 - \nu_s} \Delta\sigma_z' \tag{4–45}$$

where ν_s now applies to the *soil structure*. The vertical strain, from Eq. (3–18a) and Eq. (4–45) will be

$$\Delta\epsilon_z = \frac{(1 + \nu_s)(1 - 2\nu_s)}{(1 - \nu_s)E} \Delta\sigma_z' \tag{4–46}$$

In the laboratory test this one-dimensional strain can be calculated from the initial thickness of the sample and the total compression indicated by the dial gage shown in Fig. 4–9. However, because all of the change in length has resulted from a change in the void volume, it is usual to refer, instead, to void ratios.

It is the conventional practice in running a consolidation test to apply a given load and to record the compression dial reading as it varies in time with consolidation of the soil until it becomes essentially constant, then to apply a further load, and so on. The calculated void ratios of the soil at the end of each loading stage are then plotted against the applied stresses as in Fig. 4–10. A curve AC of the shape shown in that figure is usually obtained, demonstrating that the stress-strain response of the soil is *not* linearly elastic as we have assumed. However, the portion of the curve between any two points (representing the response to one increment of stress) can be taken as straight in the analyses of the time process of consolidation.

We shall consider that, in our present test, the void ratio changes from

Fig. 4–10 *One-dimensional compressional behavior of clay soil.*

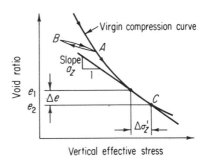

e_1 to e_2 under the application of stress $\Delta\sigma_z'$ as shown in Fig. 4–10. If this portion of the curve is taken to be a straight line, the relation between e and σ_z' is expressed by

$$\Delta e = a_z \, \Delta\sigma_z' \tag{4-47}$$

where a_z, the coefficient of compressibility, is a constant representing the slope of the line for the stress increment considered; its value will be related below to the other parameters determined previously. The volumetric strain (compression is taken as positive) of the sample as a result of the one-dimensional compression is

$$\Delta d_s = \frac{\Delta e}{1 + e_1} = \frac{a_z \, \Delta\sigma_z'}{1 + e_1} \tag{4-48}$$

Remembering that $\Delta\sigma_z' = \Delta u$ for a saturated soil, we find, for unit cross-sectional area and element thickness Δz, that the rate of change of volume R is given by the equation

$$R = \frac{a_z}{1 + e_1} \frac{\Delta u}{\Delta t} \, \Delta z \tag{4-49}$$

when the total stress remains constant with time. In this form the constants a_z and $1 + e_1$ for any increment of applied stress can be obtained from the plot of void ratio e versus effective stress as it is usually obtained. Putting Eq. (4–49) on the right-hand side of the one-dimensional flow equation gives the difference equation

$$\frac{c_z}{(\Delta z)^2} (u_2 + u_4 - 2u_0) = \frac{\Delta u}{\Delta t} \tag{4-50}$$

or in the derivative form

$$c_z \frac{\partial^2 u}{\partial z^2} = \frac{\partial u}{\partial t} \tag{4-51}$$

as before, except that c_z is now given by the equation

$$c_z = \frac{k_z(1 + e_1)}{\gamma_w a_z} \tag{4-52}$$

4-9 ULTIMATE SETTLEMENT

When the excess pore-water pressure has decreased to zero, the applied stress is balanced by an equal effective stress in the soil structure and the soil sample has changed in thickness or settled by an amount ρ, which is given by the vertical strain times the thickness $2d$ of the sample:

$$\rho = \Delta\epsilon_z \, 2d = \frac{\Delta e}{1 + e_1} \, 2d \tag{4-53}$$

Substituting Eq. (4–46) or (4–47) in Eq. (4–53), we get

$$\rho = \frac{(1 + \nu_s)(1 - 2\nu_s)}{(1 - \nu_s)E} \, \Delta\sigma_z' 2d \quad \text{or} \quad \rho = \frac{a_z \, \Delta\sigma_z' \, 2d}{1 + e_1} \quad (4\text{--}54)$$

When the effective-stress increment is equal to the increase in applied stress,

$$\rho = \frac{2d \, a_z \, \Delta\sigma_z}{1 + e_1} \quad (4\text{--}55)$$

Equation (4–54) is valid, in general, for any element of soil of given thickness Δz at any stage in consolidation, whether it is saturated or not, so that the incremental settlement $\Delta\rho$ during consolidation is given by the equation

$$\Delta\rho = \frac{a_z \, \Delta\sigma_s' \, \Delta z}{1 + e_1} \quad (4\text{--}56)$$

when $\Delta\sigma_z'$ is the increase in effective stress at that stage. This gives us a method, as we shall see below, whereby the time variation of settlement of a soil mass can be calculated.

4–10 NUMERICAL EXAMPLE

We are now in a position to study the consolidation process in the one-dimensional consolidation test. This involves the gradual dissipation of pore pressures with a simultaneous increase in vertical effective stresses leading to settlement or decrease in the sample thickness.

The initial excess pore pressure is the same everywhere in a saturated soil sample, and it is equal to the applied stress; for simplicity in a demonstration calculation, we can take the applied stress to be 100. In our example the applied stress will be taken to be constant in time, and therefore if at some stage in the calculations the excess pore pressure has decreased, say to the value 85 at a given point, we calculate by the effective-stress equations (3–52) that the effective stress there has reached the value 15. The sum of excess pore pressure and effective stress must always be equal to the applied stress.

We shall study consolidation numerically first, using Eq. (4–50), since the steps in the calculation clearly illustrate the physical process occurring. Later the exact analytical solution of the differential equation (4–51) will be given for comparison.[3]

In Fig. 4–11 a vertical cross section has been drawn through the consolidation sample of Fig. 4–9. Because drainage can only take place vertically to the porous stones top and bottom, all such vertical cross sections will possess the same excess pore pressure distribution during the dissipation process. In order to use Eq. (4–50) developed in terms of pressure differences, we must divide the sample into elemental layers of soil of thickness Δz as shown in Fig. 4–11 (the layers next to the surface have half the thickness of the others). We are going to assume that the pore pressure calculated at the center of each layer is an average value of the pore pressure throughout the layer.

The time axis is taken horizontally in Fig. 4–11, and a series of vertical lines as shown represents successive times 0, Δt, $2\Delta t$, $3\Delta t$, ... We shall calculate the pore pressures at these times at the centers of the soil elements shown. Initially, the excess pore pressure is everywhere 100, but since the porous stones have a high permeability, the excess pore-water pressure at the two surfaces of the soil specimen becomes zero immediately after application of the load $\Delta\sigma_z$. This boundary condition, which remains constant throughout the process, provides the pore pressure difference required to cause flow and consolidation. Writing Eq. (4–50) in the form

$$\Delta u = \frac{c_z \,\Delta t}{(\Delta z)^2}\,(u_2 + u_4 - 2u_0) \tag{4–57}$$

for the point-numbering convention shown in Fig. 4–11, it can be seen that the change in excess pore pressure Δu over the time interval Δt can be calculated from a known coefficient of consolidation c_z and the conditions of Δt and Δz to be employed in the problem. We observe at this stage, before beginning the calculation, that the parameter M, where

$$M = \frac{c_z \,\Delta t}{(\Delta z)^2} \tag{4–58}$$

is dimensionless. Now in numerical analysis in general, the numbers we employ are never exactly correct, but always contain a small error. Difference equations such as Eq. (4–58) above are also never exactly right, but may approach the correct expression (the derivative form) as Δt and Δz become very small. Consequently, at any stage in the calculations, we perform an approximate computation using numbers which are not quite correct to begin with. In circumstances such as this, it is desirable to

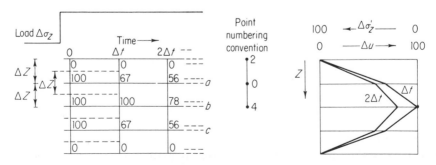

Fig. 4–11   *One-dimensional numerical solution steps.*

choose a numerical procedure which tends to diminish the numerical errors.

In the present circumstances it has been found that the numerical method represented by Eq. (4–57) is valid and converges to the right solution only if a value of the operator M is chosen less than $\frac{1}{2}$. The process is then called *stable*, and for practical reasons of computation values of $\frac{1}{3}$ or $\frac{1}{4}$ are frequently employed. For our problem we shall choose M to be equal to $\frac{1}{3}$, and thus for a given soil, c_z, sample thickness, and convenient Δz, Δt is determined by Eq. (4–58).

Putting $M = \frac{1}{3}$ in Eq. (4–57) gives a special form of that equation which is convenient for computation,

$$\Delta u = \tfrac{1}{3}(u_2 + u_4) - \tfrac{2}{3}u_0 \qquad (4\text{--}59)$$

Alternatively, we can consider Δu as the difference between the value of u at the point zero and time t, $u_{0,t}$, and the value at the same point at time $t + \Delta t$, $u_{0,t+\Delta t}$, as follows:

$$\Delta u = u_{0,t+\Delta t} - u_{0,t} \qquad (4\text{--}60)$$

Substituting this expression in Eq. (4–59) gives

$$u_{0,t+\Delta t} = \tfrac{1}{3}(u_{2,t} + u_{0,t} + u_{4,t}) \qquad (4\text{--}61)$$

in which the subscripts indicate both the position and time at which the pore pressure is measured. This value of M therefore gives a "new" value at a given point which is the average of the "old" values at the point and the two adjacent points. We may now proceed with the calculation shown in Fig. 4–11, taking each point in turn down from the surface to be the point 0. The point at the surface, of course, stays at the value zero throughout. At point a, the pore pressure at time Δt is equal, from Eq. (4–62), to $\frac{1}{3}(0 + 100 + 100) = 67$; at b the pressure is equal to $\frac{1}{3}(100 + 100 + 100)$ and therefore remains unchanged. The calculation for point c turns out to be the same as that at point a, and the value at c at time Δt is therefore also 67. This similarity arises from the symmetry of the sample and the network of points selected, and it holds throughout the calculation. The computation proceeds along these lines until the values have been progressively reduced essentially to zero.

At the right-hand side of Fig. 4–11 a graph has been plotted to show the successive values of excess pore pressure u as a function of depth in the sample, and this figure also gives the complementary stresses $\Delta\sigma_z'$ [equal to $(100 - u)$]. The progressive decrease in excess pore pressure can be clearly seen in this figure.

The values of $\Delta\sigma_z'$ at any time computed from the excess pore pressure for each soil layer or incremental depth in the sample can be employed in Eq. (4–56) to give a method of computing the settlement which has taken place up to that time in the soil sample. Because the effective stress is also considered as an average over each elemental thickness, we can see from Eq. (4–56) that at any time $n\,\Delta t$ the total expression $\rho_{n\Delta t}$ will be equal to $\Sigma\,\Delta\rho$, where

$$\rho_{n\Delta t} = \sum \Delta\rho = 2\,\frac{a_z\,(\Delta z/2)\,\Delta\sigma'_{z0}}{1+e_1} + 2\,\frac{a_z\,\Delta z\,\Delta\sigma'_{za}}{1+e_1} + \frac{a_z\,\Delta z\,\Delta\sigma'_{zb}}{1+e_1}$$

$$(4\text{–}62)$$

in which $\Delta\sigma'_{z0}$, $\Delta\sigma'_{za}$, and $\Delta\sigma'_{zb}$ are the effective-stress increments at the surface and points a and b, respectively, at time $n\,\Delta t$. At the surfaces $\Delta\sigma'_{z0}$ is always equal to 100. The figure 2 in front of the first two terms takes account of the symmetry of the system, in that conditions at the surfaces and at points a and c in Fig. 4–11 are always identical at the same time. Collecting terms together in Eq. (4–62) gives

$$\rho_{n\Delta t} = \frac{a_z\,\Delta z}{1+e_1}\,(\Delta\sigma'_{z0} + 2\Delta\sigma'_{za} + \Delta\sigma'_{zb})$$

$$(4\text{–}63)$$

It is convenient for many purposes to express the settlement at any time in terms of the ultimate settlement to be expected, and this can be done by dividing both sides of Eq. (4–63) by both sides of Eq. (4–55) to define a new term U, the *average degree of consolidation* of the sample

$$U = \frac{\rho_{n\Delta t}}{\rho} = \frac{\Delta z}{2d}\,\frac{\Delta\sigma'_z + 2\Delta\sigma'_{za,n\,\Delta t} + \Delta\sigma'_{zb,n\,\Delta t}}{\Delta\sigma'_z}$$

$$(4\text{–}64)$$

It will be seen that Eq. (4–64) can be generalized for any problem to the equation

$$U = \frac{\Sigma\,\Delta\sigma'_{zm,t}\,\Delta z}{\Delta\sigma'_z\,2d}$$

$$(4\text{–}65)$$

or

$$U = \frac{\Sigma\,U_{m,t}\,\Delta z}{2d}$$

$$(4\text{–}66)$$

where $U_{m,t}$ is the degree of consolidation at point m and time t.

In the particular case of Fig. 4–11 we can calculate the settlement at time $2\Delta t$, for example, as follows:

$$\rho_{2\Delta t} = \frac{a_z\,\Delta z}{1+e_1}\,(100 + 2\cdot 44 + 22)$$

$$= \frac{a_z\,\Delta z}{1+e_1}\,(210)$$

since the effective stresses are equal to 100 minus the pore pressures calculated in Fig. 4–11. However, the ultimate settlement is

$$\rho_u = \frac{a_z\,\Delta z}{1+e_1}\,(400)$$

so that the dimensionless degree of consolidation at this stage of the process is

$$U = \frac{210}{400} = 0.525$$

This value can be calculated for all times during consolidation to give a curve of degree of consolidation versus time. The fact that we have been able to compute this number without reference to numerical values of sample thickness, pressures, or soil properties indicates that the calculation applies to one-dimensional consolidation in a layer of soil of any thickness drained at top and bottom and subjected suddenly to an applied load which does not vary in time after its application.

When a dimensional analysis of consolidation is performed or the consolidation equation is reduced to dimensionless form, a convenient dimensionless time factor T can be defined to describe the time variation in the consolidation process in a manner similar to that in which U describes the state of consolidation without regard to the dimensions of the system. The time factor is given by the equation

$$T = \frac{c_z t}{d^2} \tag{4-67}$$

From this equation we can see that

$$\Delta T = \frac{c_z \, \Delta t}{d^2} \tag{4-68}$$

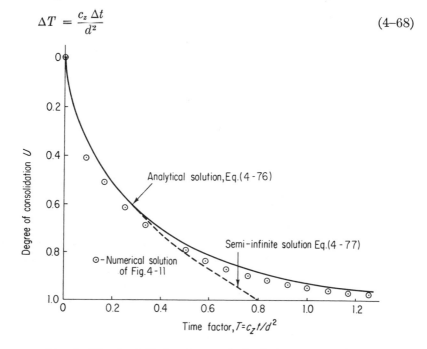

FIG. 4–12 *Analytical and numerical solutions compared.*

However, ΔT can be obtained from Eq. (4–58) and substituted in Eq. (4–68) to give

$$\Delta T = M \left(\frac{\Delta z}{d} \right)^2 \tag{4-69}$$

Here, $\Delta z/d$ is a dimensionless number giving the ratio of the elemental thickness to the half depth of the layer or sample. For the example of Fig. 4–11 we see that

$$\Delta T = \tfrac{1}{3}(\tfrac{1}{2})^2 \approx 0.083$$

The dimensionless plot of the numerical calculation of Fig. 4–11 is shown in Fig. 4–12.

4–11 ANALYTICAL SOLUTION

The analytical solution to the above problem described by Eq. (4–51) with the given boundary conditions has also been obtained; in terms of excess pore pressures, it is

$$1 - U_{Z,T} = \frac{\Delta u}{\Delta \sigma_z} = \frac{4}{\pi} \sum_{m=0}^{\infty} \frac{1}{2m+1} \left[\exp \frac{-\pi^2 (2m+1)^2 T}{4} \right. \\ \left. \sin \frac{(2m+1)\pi Z}{2} \right] \tag{4-70}$$

where m is an integer and $Z = z/d$, T is given by Eq. (4–67), and $U_{Z,T}$ is shown in Fig. 4–13 as a family of curves analogous to the top half of the diagram in the right of Fig. 4–11.

From the preceding numerical analysis, the degree of pore pressure dis-

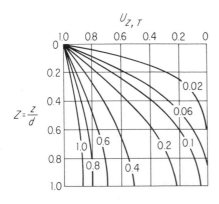

FIG. 4–13   *Analytical solution to one-dimensional consolidation equation for single soil layer. Values of dimensionless time T are shown on the curves.*

sipation and the degree of consolidation are related in the following way:

$$U_{m,T} = 1 - \frac{u_{m,T}}{\Delta\sigma_z} \tag{4-71}$$

or

$$U_T = 1 - \frac{\bar{u}_T}{\Delta\sigma_z} \tag{4-72}$$

where $\bar{u}_T$ is the space average of excess pore pressure at time T. The derivative form of the average pore pressure is

$$\bar{u}_T = \frac{1}{d} \int_0^d u_{m,T} \, dz \tag{4-73}$$

(since the process is symmetrical about the center of the sample) and of the average effective stress is

$$\Delta\sigma'_{zT} = \frac{1}{d} \int_0^d \Delta\sigma'_{zm,T} \, dz \tag{4-74}$$

The average degree of consolidation analogous to Eq. (4-66) is therefore

$$U_T = \frac{1}{d} \int_0^d U_{m,T} \, dz \tag{4-75}$$

From the analytical solution equation (4-70) and Eqs. (4-72) and (4-73), the analytical solution for the average degree of consolidation as a function of dimensionless time is

$$U_T = 1 - \frac{8}{\pi^2} \sum_{m=0}^{\infty} \frac{1}{(2m+1)^2} \exp \frac{-\pi^2(2m+1)^2 T}{4} \tag{4-76}$$

This solution has been plotted in Fig. 4-12 for comparison with the numerical solution obtained previously.

Another, simpler analytical solution of interest may be mentioned here, since it is frequently of value in approximate calculations. It is the solution to the one-dimensional problem of the consolidation of a layer of compressible soil, as shown in Fig. 4-14, extending to infinite depth, drained only at the ground surface, and subjected to applied stress at the surface as before. Only the solution for the average degree of con-

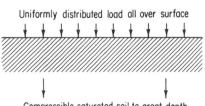

FIG. 4-14   *One-dimensional consolidation of semi-infinite compressible soil.*

solidation U_T will be given:

$$U_T = 2\sqrt{\frac{T}{\pi}} \tag{4-77}$$

and this is shown plotted in Fig. 4–12 for comparison with the case of consolidation in a layer of finite depth. It is seen that it corresponds very closely with the layer analytical solution up to about $U = 0.5$. There is, of course, no limit to the ultimate settlement of a layer of infinite thickness. The reason for the early identity of the two curves is simply that the diminishing pore pressures at the surface of the finite layer are not aware that the layer is limited in extent until the influence of the other surface makes itself felt.

4-12 DETERMINATION OF COEFFICIENT OF CONSOLIDATION

Because of the parabolic relationship between U and T, expressed by Eq. (4–77), a plot of U versus $\sqrt{T}$ for the *finite* layer starts out as a straight line before gradually curving to become asymptotic to the $U = 1.0$ abscissa. This fact is usefully employed in one method of determining the coefficient of consolidation of a compressible soil from a laboratory consolidation test.

Although the coefficient of consolidation is given by Eq. (4–52), it is not usually accurate or convenient to establish a value of the coefficient by obtaining values of k, a_z, and e from different tests for substitution in the equation. Instead, since the coefficient directly affects the time-settlement curve of a given soil when plotted in dimensional terms, such a curve forms the best way of estimating the coefficient of consolidation. The curve of U versus $\sqrt{T}$ from the analytical solution equation (4–76) is plotted in Fig. 4–15a alongside a curve plotted from a typical laboratory consolidation test on a clay sample of thickness 0.75 in., Fig. 4–15b.

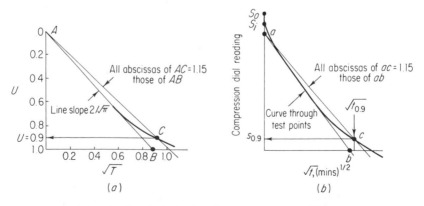

FIG. 4–15   *Calculation of consolidation coefficient.*

The latter curve is obtained from readings of the compression dial (Fig. 4–9) at various times after the application of any load increment. Typically the laboratory curve demonstrates an initial compression from reading s_0 to s_i and then resembles the theoretical solution for some distance, but it tails off gradually into a line with no clear-cut horizontal asymptote. This behavior renders identification with the theoretical solution difficult, and therefore the following procedure has been used.

A straight line ab is fitted to the nearly straight portion of the laboratory curve as well as possible to intersect the zero time axis in a. This line is equivalent to AB, the straight line of slope $2/\sqrt{\pi}$ in Fig. 4–15a. Since the line AB intersects the $U = 1.0$ line at $\sqrt{T} = \sqrt{\pi}/2 = 0.88$ in the theoretical curve in Fig. 4–15a, we could calculate c_z for our test sample if there was a definite intersection of the line ab with an ultimate settlement or compression abscissa in Fig. 4–15b at a calculable time. Usually there is not one, so that another suitable point of similarity on the two curves must be identified. The calculation is made the more accurate the nearer this point is to complete consolidation, and in practice $U = 0.9$ is usually selected. The point C on the theoretical curve corresponding to $U = 0.9$ is marked, and it is found from Eq. (4–76) that $T = 0.848$ corresponds to this point. Besides the settlement curve itself, another line is needed to mark point C; it is obtained by drawing a straight line AC. It is found from the theoretical results that the abscissas of AC are equal to 1.15 times those of AB. The locating of the corresponding point on the experimental curve of Fig. 4–15b is therefore carried out by drawing a line ac through a with abscissas 1.15 times those of ab. The line ac intersects the plotted curve in c, whose ordinate is taken as representing a state of settlement equivalent to 0.9 times the theoretically final value and whose abscissa indicates the corresponding real time ($\sqrt{t_{0.9}}$ on the square-root scale) at which the settlement or compression reaches that amount.

From the theoretical curve we know the dimensionless time factor at this point, and we can therefore set up the equation

$$T_{0.9} = 0.848 = \frac{c_z t_{0.9}}{d^2}$$

The time $t_{0.9}$ is obtained from the abscissa of point c on the experimental curve of Fig. 4–15b; the thickness of the sample is known; and thus c_z can be calculated. It is usual to compute c_z in this (or another) way for each increment of applied stress in a consolidation test, and the values will generally be found to vary. The variation occurs because the void ratio and structure of the soil, and therefore its permeability and compressibility, vary with increasing applied stress in a nonlinear way as shown in Fig. 4–10.

4-13 BEHAVIOR OF REAL SOILS

If in the laboratory a clay soil is mixed with water to a consistency close to the liquid limit and placed in the consolidation apparatus of Fig. 4–9, it will exhibit, under gradually increasing stresses, the consolidation behavior shown by the curve marked "virgin compression curve" in Fig. 4–10. This curve might represent the loading history of a clay soil as it is deposited in a lake or in the sea as a result of sedimentation of clay particles. Each element of clay becomes buried to a successively greater depth and is subjected to a larger vertical stress with time. If it has never been subjected to a higher stress, the soil is termed *normally consolidated*. If at a point A in Fig. 4–10 the stress is removed, it is found that the soil does not expand back along the virgin compression curve, but instead expands along a curve of much flatter slope AB, and the soil does not therefore behave elastically. Along the path AB the soil is said to be *overconsolidated*. On recompression, another, slightly different curve is obtained following generally along the path from B back to A again, and at point A the soil resumes the virgin compression behavior on increase of stress.

Since the process of loading and unloading represents approximately the stress history of a sample of soil removed from the natural ground and stored in a laboratory for some period before testing, we generally find that a natural soil specimen exhibits a void ratio versus vertical effective stress curve of the shape BAC in Fig. 4–10. The maximum curvature on the diagram occurs at point A and indicates approximately the magnitude of the previous maximum vertical effective stress applied to the soil. This may be higher than the effective stress calculated to exist at the level of the sample at the time of extraction for reasons discussed below.

Because it is frequently of interest in practice to determine the maximum stress, a construction which is shown in Fig. 4–16 has been suggested. There the $e - \sigma_z'$ curve is plotted by using a logarithmic stress scale, whereupon a straight line is generally found to represent the virgin compression, or normally consolidated, portion of the curve. In the construction, this straight line is produced in the direction of higher void

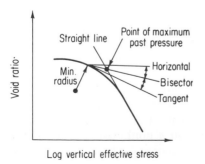

Fig. 4–16 *Determination of maximum past pressure.*

ratios. At the point of maximum curvature of the curve a tangent is drawn to it and also a horizontal line as shown in the figure. The angle between the tangent and the horizontal is bisected, and the point of intersection of the bisector with the produced straight-line portion of the original curve is located. This point is considered to represent the void ratio and effective stress (called the *preconsolidation pressure*) of the material at its previous maximum state of compression.

The procedure is especially valuable if the natural soil in the ground, as is often the case, has been subjected to higher loads in its geological history than those existing at the time the soil sample was obtained; these higher loads can have developed as a result of desiccation or from the presence of an ice sheet during a preceding period of glaciation. In some areas the thickness of the ice sheet has been calculated from the preconsolidation pressure determined by the above procedure.

In our analysis of consolidation we assumed that for any increment of applied vertical stress the relationship between void ratio and applied stress could be assumed to be linear in figures such as Fig. 4–10; this describes the behavior of the soil adequately for small increments of applied stress on either normally consolidated or overconsolidated soils so long as the soil is subjected only to increasing stresses. However, the application and removal of a stress to the surface of a normally consolidated soil will cause first a consolidation behavior represented by points along the slope AC in Fig. 4–10 and next an expansion or rebound along curves generally parallel to AB. The slope of the curve AB is different from that of AC, and therefore the linear behavior, which we assumed in our previous analysis of consolidation, holds in practice only so long as an increase in load occurs, and our calculations of settlement and pore pressure change are valid only if no swelling or rebound takes place anywhere in the soil under stress. The behavior of overconsolidated soils for stresses less than the maximum past pressure is more nearly elastic.

In general during construction, only increasing loads are applied to soil, but in certain special cases, such as oil storage tanks, reservoirs, and other structures loaded by fluids, the level of the fluid may fluctuate from time to time and, indeed, the fluid may on some occasions be removed entirely, so that the soil is subjected to cyclically increasing and decreasing stresses. The analysis of consolidation or settlements (in both the positive and negative senses) in such circumstances requires a more complex analysis than the one given here.

Another aspect of real soil behavior must be noted at this point. In Fig. 4–15b were plotted the results obtained from a consolidation test in terms of compression dial reading versus the square root of time. This curve is shown also in Fig. 4–17a with a linear time scale and in Fig. 4–17b versus the logarithm of time. In all of the curves it will be observed that the dial reading does not tend toward a limiting value at long times, and this leads, as mentioned before, to the difficulty of uniquely determining when the sample is 100 percent consolidated. What appears is a deformation which steadily increases, even at very long times, as shown by the logarithmic plot of Fig. 4–17*b*.

In our analysis of consolidation we assumed that the soil structural skeleton exhibited an elastic behavior and was capable of withstanding the applied stress without further compression after the pore pressures had finally dissipated. In fact, the structure of fine-grained soils, as described in Chapter 1, is such that there is no final structural position at which the arrangement of grains is able to absorb the effective stress. Instead, the structural skeleton continues to deform slowly owing to slippages and rotations of the individual particles under local stresses even when pore-water pressures have completely dissipated. Thus, further distortion and compression of the sample occurs at essentially zero pore pressures for long times after the application of stress. The structural skeleton of the soil can be said to possess a high *viscosity,* so that *flow* takes place under effective stresses. The phenomenon that we have described is called *secondary compression* or *creep.* It is more important in some kinds of soils than others, and only rarely is it taken into account in actual design or analysis procedures. There is at present no generally accepted formal way of accounting for creep behavior, although various theoretical solutions for consolidation including structural viscosity have been discussed in the literature.

In most situations in practice where consolidation must be considered, the distribution of load on the ground surface and the thickness of the compressible layer are such that, strictly speaking, a two- or three-dimensional consolidation analysis should be carried out. In most cases it is not carried out because of the complexity and difficulty of such an analysis, together with the difficulty of obtaining sufficiently accurate soil

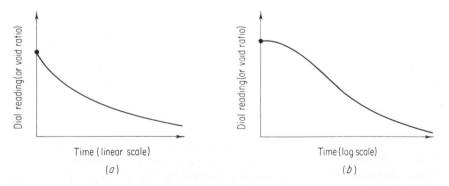

FIG. 4–17   *Creep or secondary compression in clay soils. (a) Time on a linear scale. (b) Time on a logarithmic scale.*

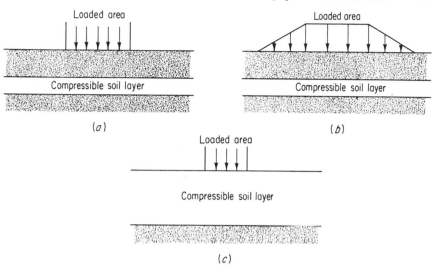

FIG. 4-18 *Practical consolidation conditions. (a) Building on a mat foundation. (b) Highway fill. (c) Thickness of compressible layer, large compared with building dimensions.*

information to justify the labor of performing it. It is usual in situations such as represented by Fig. 4-18a and b, respectively showing a loaded building on a mat foundation and fill on the ground surface, to calculate the vertical effective stresses in the compressible layer below the center and edges of the structure by means of an elastic-stress analysis such as given in Chapter 3. The assumption is then made that the compressible layer consolidates as a result of one-dimensional drainage to either or both of its top and bottom surfaces depending on the drainage conditions. When the compressible layer is fairly thin with respect to the dimensions of the loaded area, a result adequate for engineering purposes may be obtained. However, in cases such as shown in Fig. 4-18c, where the compressible layer is of considerable extent compared to the load, less approximate methods are required. For these situations some solutions for the three-dimensional consolidation process have been presented in the literature, usually in the form of charts or diagrams giving the settlement under a corner of the structure as a function of time.

4-14 EXAMPLE OF SETTLEMENT CALCULATION

When, as in Fig. 4-19a, a building or fill exerts stresses on the surface of the ground over an area which is large with respect to the thickness of the consolidating layer and its depth below the surface, a one-dimensional consolidation and settlement analysis can proceed as follows. Samples of the compressible layer are obtained and subjected to con-

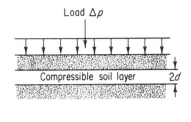

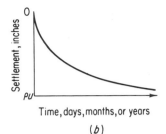

(a)   (b)

FIG. 4–19   *Settlement estimate. (a) Soil profile. (b) Predicted surface settlement versus time.*

solidation tests in the laboratory over a stress range which includes the ambient stress and the increment of applied pressure Δp to be added in the field. The compressibility coefficient a_z of the soil sample is obtained from the void ratio versus effective-stress curve as the slope of the curve over the stress increment Δp. The coefficient of consolidation c_z of the compressible layer is determined for the Δp increment as described previously. These properties are usually obtained as averages for several samples.

Knowing the initial void ratio of the compressible layer, its compressibility a_z, and its thickness $2d$, the ultimate settlement under an effective stress of Δp can be calculated from Eq. (4–55). Then the ordinate axis of a diagram such as Fig. 4–19b can be set up on the basis of this ultimate settlement; this axis corresponds to the U axis of Fig. 4–12. In Fig. 4–12 the horizontal axis T is a dimensionless time equal to $c_z t / d^2$. Since we have obtained the coefficient of consolidation c_z from the test and the thickness of the layer is known, we can calculate the value of time t in the field which corresponds to any value of T and therefore, through Eq. (4–76), to U_T. The theoretical one-dimensional degree of consolidation versus time curve of Fig. 4–12 is thereby converted to give in Fig. 4–19b a settlement versus time curve for the real soil under the applied-stress increment Δp.

PROBLEMS

4–1   Mark the boundary flow and equipotential lines clearly in Fig. 4–1a assuming that a section of a rectangular tube is shown. Draw the flow net in the soil.

4–2   Calculate the quantity q/kH in Eq. (4–19) for Fig. 4–1a from your flow net. Draw a flow net with a different b/a ratio and check that you get the same q/kH quantity.

4–3   Practice sketching a flow net in the configuration of Fig. 4–1b following the method shown in Fig. 4–4.

4 - 4 Change the depth of penetration of the wall of Fig. 4–4 to be three-quarters of the way to bedrock and sketch the new flow net.

4 - 5 Calculate q/kH for your nets of Probs. 4–3 and 4–4.

4 - 6 Plot q/kH versus relative depth of penetration of the wall in Fig. 4–1b. Does the extra driving distance of the sheet piles reduce the flow proportionately?

4 - 7 If CG is the reference datum and the figure is to scale with $H = 20$ ft, calculate the total, elevation, and pressure heads at points B and J in Fig. 4–1b (after drawing the flow net).   *Answer*   10, −20, 30

4 - 8 What is the head gradient across the flow rectangle in Fig. 4–3b if the thickness of the soil layer in Fig. 4–3a is 20 ft and $H = 30$ ft? What are the vertical gradients at points C, K, L, and N?

Answer   1.1, 1.07, 0.86, 0.57, 0.20

4 - 9 Draw a sketch like Fig. 4–5a with the angle $\alpha_1 = 45°$ and plot the emerging flow line angle α_2 for ratios of k_1/k_2 equal to 10, 2, 1, 0.5, and 0.1. What would you expect the angle to be if the upper material is one soil in Table 2-2 and the lower material is the next finer class of material? What do you conclude from this?

4 - 10 From the flow net you drew for Prob. 4–3 pick off values of total head at a few of the grid points shown in Fig. 4–6 and check that Eq. (4–22) holds at these points. If it does not, use Eq. (4–22) successively to calculate new values of head at the grid points.

4 - 11 The soil in Prob. 4–8 has a specific gravity G_s of 2.6 and a void ratio of 0.7. Is the soil to the right of the wall stable?

4 - 12 Check the gradients and stability of the wall in your solution to Prob. 4–4. What do you conclude?

4 - 13 Sketch the flow net for Fig. 4–8c, bearing in mind the equipotential requirement of Fig. 4–8a. Calculate q/kH for the dam from your flow net and compare with the result given by Eq. (4–29).

4 - 14 A saturated clay soil sample had an initial thickness of 0.75 in., a diameter of 3 in., wet weight of 170.0 grams, dry weight of 130.7 grams, and specific gravity of 2.75. Calculate its void ratio versus pressure curve on both linear and logarithmic scales of pressure for the following final dial readings during a consolidation test.

Pressure, tons/ft²	0.25	0.5	1.0	2.0	4.0	8.0	16.0
Final dial reading, in.	0.00	0.001	0.002	0.008	0.027	0.052	0.070

4 - 15 What was the maximum past pressure to which the soil of Prob. 4–14 had been subjected?   *Answer*   2.2 tons/ft²

4 - 16 If the soil sample of Prob. 4–14 was taken from a depth of 25 ft below existing ground surface, was it overconsolidated at that depth?

4 - 17 From the void ratio versus linear pressure curve of Prob. 4–14 calculate the coefficient of compressibility a_z over (a) the pressure range 0.5 to 1.0 tons/ft² and (b) the range 4 to 8 tons/ft².

Answer   (a) 0.01 (tons/ft²)⁻¹; (b) 0.015 (tons/ft²)⁻¹

4-18   The soil, of which Prob. 4–14 gave typical consolidation test results, forms a 20-ft-thick layer in the ground extending down from a depth of 15 ft. The water table is also at the 15-ft depth, and a sand soil of total unit weight 100 pcf exists both above and below the clay soil layer. It is proposed to apply a fill all over the ground surface, which will subject the surface to a uniform load of 1 ton/ft². What will be the ultimate settlement under this fill due to the clay layer alone?   *Answer*   1.72 in.

4-19   If the coefficient of consolidation c_z for the soil of Prob. 4–18 is 0.1 ft²/day, how long will it take for one-half of the ultimate settlement to occur?   *Answer*   200 days

4-20   Plot the settlement versus time curve for the situation of Prob. 4–18.

4-21   Carry out a numerical consolidation calculation for the case of Prob. 4–18. Use real pressures and times.

4-22   What is the excess pore pressure at the center of the clay layer of Prob. 4–18 when one-half of the total settlement has occurred?   *Answer*   1,540 psf

4-23   The accompanying table gives the dial readings recorded versus time in a laboratory consolidation test on a 0.75-in.-thick sample after application of a load increment. Plot the compression dial reading versus the square root of time and determine the coefficient of consolidation of the soil for a doubly drained sample.   *Answer*   0.28 ft²/day

Time, min	0.1	0.2	0.4	0.6	1	2	3	4	6
Dial reading, in.	0.008	0.012	0.016	0.020	0.025	0.035	0.041	0.044	0.048

REFERENCES

1. D. W. TAYLOR, "Fundamentals of Soil Mechanics," John Wiley & Sons, Inc., New York, 1948.
2. M. E. HARR, "Groundwater and Seepage," McGraw-Hill Book Company, New York, 1962.
3. R. F. SCOTT, "Principles of Soil Mechanics," Addison-Wesley Publishing Company, Inc., Reading, Mass., 1963.

5 | Shearing Strength of Soils

At the beginning of Chapter 3 the concept of soil deformation taking place under an increasing load at the ground surface and leading eventually to failure of the ground underneath the load was introduced and discussed with respect to Fig. 3–1. This behavior occurs in the mass as a result of a similar behavior in the individual elements of soil composing the mass. We are now going to examine the limiting stress conditions that cause failure in samples of different soils when the sample or element is small enough or the loading is so arranged that the same stress conditions exist throughout the entire element. This is called a *homogeneous stress state*, and it does not occur, for example, in the soil stressed by a loaded footing at the surface.

In Chapter 3 we saw that it was convenient to separate the behavior of materials into two areas for study: (1) the volume change taking place under hydrostatic stresses and (2) the shearing strain or distortion of the material under shearing stresses. This consideration is also important here, since the behavior represented by Fig. 3–1 cannot occur in a metal as a result of increasing compressive *hydrostatic* stresses only. The material will continue to deform without increase in the stresses only when shearing stresses are applied. It is also found in metals that the shearing stress which will cause unlimited deformation is unaffected by the level of hydrostatic stress existing in the material in the stress range of interest to engineers.

To some extent the same considerations apply to soil. Under an increasing hydrostatic stress the volumetric behavior of any soil can be represented by the compression curve of figures such as Fig. 4–10 when an appropriate scale of void ratio is chosen for the soil being studied. For sand, the void ratio will not change by very much up to stresses at which the grains begin to crush (>100 psi), whereas for clays, void ratio changes of 100 percent or more are possible as a result of applied effective stresses in the range of engineering interest. Clays and sands differ in their behavior when some of the stress is removed, but we are not concerned with that here. It will be noticed in Fig. 4–10 that there is no stress at which an unrestrained volume decrease, corresponding to "failure," occurs. The soil only gets stiffer as the stress increases.

As with metals, the situation changes when shearing stresses are applied to soils. Figure 5–1 has been drawn for a soil which has been subjected first to some applied hydrostatic effective stress and then to shearing stresses at constant hydrostatic effective stress. The deformation or shearing strain increases as the shearing stress increases, and it is found that a limiting shearing stress, at which the shearing distortion continues without limit, will be reached. In this respect various soils exhibit different characteristics represented by curves *a* and *b* of Fig. 5–1; in general, loose to medium dense sands and normally consolidated clays exhibit shearing stress versus deformation curves of type *a*, whereas dense sands and overconsolidated clays behave as shown by curve *b* of Fig. 5–1. The two curves are frequently referred to as representing stable (*a*) and unstable (*b*) behavior, respectively.

An important distinction must be made between soils and metals at this point. When a metal is subjected to shearing stress and deforms in a manner similar to that shown by one of the curves of Fig. 5–1, a corresponding measurement of the volume of the metal or its density shows no change as shearing progresses if the hydrostatic stress remains unchanged; the volumetric strain is determined by the hydrostatic stress alone, according to Eq. (3–25). However, when soils are tested in shear at constant hydrostatic stress, a volume change *does* manifest itself, and this must be taken into account in an assessment of the material behavior. Because of this phenomenon, it is usual to accompany the curves of Fig. 5–1 by the corresponding diagram underneath, showing the volume increase or decrease during the shearing test.

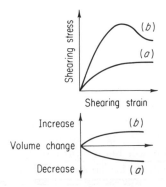

Fig. 5–1   *Behavior of soil under shearing stress.*

Most frequently, curves of the shape of *a* are accompanied by volume decrease as shearing or deformation progresses, and curves *b* are accompanied by a volume increase during shear. The reason for this behavior is that the structural arrangement of soil particles, whether in sands or clays, is important in the resistance of the soil to both applied shearing and hydrostatic stresses. As a result of the application of shearing stresses, the structure of the soil changes as the soil grains displace and are rotated. In a dense granular soil the material tends to expand during shear because the grains have to climb up over each other to permit distortion to occur.

Analogously, but with a different mechanism, shearing of a highly over-consolidated clay also gives rise to a volume increase as the interparticle contacts developed by the high past pressure are disrupted by shearing. Loose sands and normally consolidated clays contract as the shearing strains break down the soil structure which withstood the hydrostatic stresses without shear. Thus, in a shearing test not only the shearing stresses but also the hydrostatic stresses do work on the material (as its volume changes) during the application of shearing stresses to a stable material, whereas during shearing of an unstable material, expanding in volume during shear, energy must be supplied both for the shearing and the dilatational work. In laboratory analyses of the shearing behavior of soils, the energy expended in this way must be taken into account.

It is observed from the structural changes discussed above that the soil at the end of a shearing stress test is, in general, not the same as that with which the test began; it will have changed in void ratio if volume change is permitted, and it will also have changed in the structural arrangement of its grains. Consequently, when we refer to the "shearing strength" of a given soil material as represented by the maximum value of shearing stress attained in diagrams such as Fig. 5–1, it is important to distinguish between the initial and final states of the material. We shall see in this chapter the importance of the volume changes which take place in all soils, particularly as a consequence of shearing.

5–1 SHEARING STRENGTH TESTS

An element of soil in ground subjected to stresses applied by some structure generally undergoes changes in its stress state in three dimensions; the difference between the old and new stress states always involves changes in both the hydrostatic and the shearing stresses which have been applied to the soil. Since the behavior of soil is very complex, even to the extent to which it is outlined in Chapters 3 and 4, in its responses to hydrostatic and to shearing stresses, it has not so far been possible in soil mechanics to introduce any theoretical concepts which enable us to predict the strength of a soil for any given change in the soil stress state from soil properties such as unit weight, water content, and grain-size distribution. It is always necessary to test the soil in the laboratory by bringing it to its "undisturbed" stress state and then subjecting it to a change in stress state as close as possible to that which it would undergo in the field under the applied loads.

However, two considerations prevent us from doing this in a completely satisfactory way: one is theoretical and the other practical. The stresses which are developed in an element of soil in the field as a result of loads applied elsewhere to the mass depend on the deformational behavior of the soil. Since we cannot correctly describe this behavior mathematically, we cannot predict the stress changes in a given element. Second, it is not a practical matter at present to test a soil sample in the laboratory under a general three-dimensional test with an arbitrary variation of stress components because of the difficulty of constructing and operating suitable apparatus.

At this time, the method employed to overcome the first of these difficulties, when it is necessary to do so, is to calculate the stress state in a soil and changes in it on the basis of the assumption that the soil behaves linearly elastically. The second problem is usually taken care of by neglecting the deformational behavior of the soil and concentrating on a determination of its shearing strength starting from the condition of, and stress state in, the undisturbed soil. This strength is then used, together with an assumption regarding the manner in which the soil can fail under the given loading conditions, in a computation of the ultimate or failure load. One of two types of test apparatus is usually used to find the strength, and both will be described.

In the early stages of the development of soil mechanics as a science, a shearing apparatus of the type illustrated in Fig. 5–2 was developed; and after a period in which its use declined, it seems to have become more popular again. The reason for its use is that it is frequently found, when a failure occurs in the field, that distinct shearing planes or surfaces have appeared in the soil. The apparatus of Fig. 5–2, called a shearing box, essentially consists of a piece of equipment similar to the consolidation apparatus of Fig. 4–9 in which provision is made for stressing the soil in a vertical direction. It may be circular or square in plan. However, in the shearing box the container holding the soil is made of two halves which may be moved horizontally with respect to each other, so that the soil is sheared along its midplane. A test consists in applying a given nor-

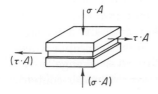

FIG. 5–2   *Direct-shear test (schematic).*

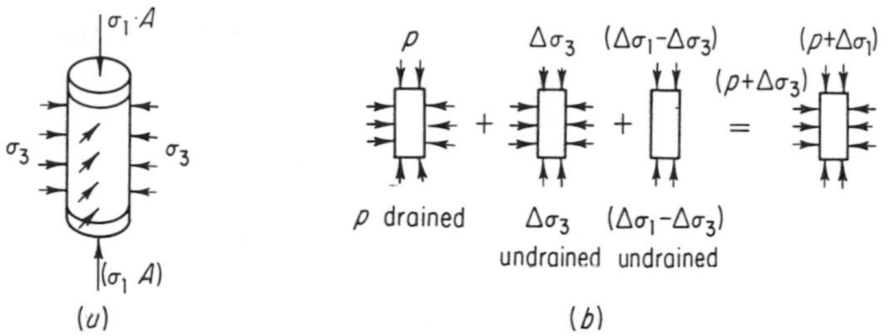

FIG. 5–3 *Triaxial test (schematic). (a) Test arrangement. (b) Stress conditions.*

mal (vertical) stress to the soil (to simulate the *in situ* conditions) and allowing the soil to consolidate or not, depending on the test. Next, the soil is sheared by displacing one-half of the box with respect to the other with a measured shearing force. Both the relative lateral displacements of the top and bottom halves and the vertical movement of the top half of the box are measured by means of dial gages. From this test, a plot of shearing stress versus lateral displacement will appear in the form of the upper diagram of Fig. 5–1, whereas a plot of the vertical versus lateral displacement gives a figure similar to the lower diagram of Fig. 5–1.

Because of concern with the nonhomogeneous stress conditions in the shear box and a desire to control the test conditions more carefully, a more complicated test was developed during the 1930s. This test endeavors to simulate more realistically the stresses acting upon the soil under various assumed conditions in the field. The apparatus, shown diagrammatically in Fig. 5–3a, is known somewhat misleadingly as a triaxial apparatus. In it the soil in the form of a vertical cylinder with length approximately twice its diameter is capped top and bottom with two relatively rigid porous stones backed by metal disks and is enclosed in a thin cylindrical rubber sheath or membrane. This test unit is placed inside a test chamber consisting of a large Lucite cylinder which is held in place by metal plates through which a loading piston and pressure lines pass. The space between the sample and the Lucite cylinder is filled with liquid to which a pressure can be applied. Consequently, as shown in Fig. 5–3a, the soil sample can be subjected to a hydrostatic stress acting on the rubber membrane and on the capping plates top and bottom. The piston passing through the top restraining plate is used to transfer a force to the capping plates on the soil cylinder to give an axial stress in addition to the hydrostatic stress acting upon the end disks. Thus the soil is subjected to an axially symmetrical stress state, and various tests of a more general nature than those of the shear box of Fig. 5–2 can be carried out on it.

The volume changes which take place in soils under both hydrostatic and shearing stresses are of importance in both the types of tests discussed above. The consequences of these will be examined in more detail in the following section.

5-2 VOLUMETRIC CHANGES; DRAINAGE AND PORE PRESSURES

In the introduction to this chapter, and also in the analyses in the last section of Chapter 3, we saw that soils in general undergo volume changes as a result of both hydrostatic and shearing stresses provided they are free to do so, that is, if drainage and volume change are permitted. In the analysis of Chapter 3 we also saw that a calculation could be made of pore-water pressure resulting from applied stresses (both hydrostatic and shearing) in a soil sample which was not permitted to undergo volume change. Because of the effective-stress principle, the volume changes imposed by drainage conditions in the soil have very important implications in both our tests and the application of test results in the field.

In both the direct-shear and the triaxial tests porous stones are provided; by means of them, the sample can be drained in the course of the test before or during shearing. In the direct-shear test, the soil is confined laterally during the preliminary application of the vertical stress as in the consolidation test. Therefore, although the initial stress state, as we saw in Chapter 4, is hydrostatic in the soil, shearing stresses will develop in the soil skeleton as drainage takes place, and in its consolidated state the lateral effective stress in the soil will be less than the vertical one. In the triaxial test it is usual, first of all, to apply a hydrostatic stress to the soil sample through the medium of the chamber liquid, and consolidation can take place in the soil by drainage through the porous stones top and bottom. In this case, however, the final stress state in the soil at the end of consolidation is still one of hydrostatic effective stress except for certain considerations (usually ignored) regarding the relative deformations of the soil near the rigid capping plates.

Therefore, if we were to test the same soil in the two apparatus in order to plot a void ratio versus effective-stress curve of the type of Fig. 4-10 and were to make the abscissa vertical effective stress for the direct-shear apparatus and hydrostatic effective stress for the triaxial apparatus, we would find that the same soil would give rise to two different curves because of the presence of shearing distortions in the direct-shear test (and in the consolidation apparatus also). These shearing distortions cause volume changes according to the considerations discussed in Chapter 3. This is a simple illustration of the care which must be taken in comparing tests and ultimately strengths developed in the two apparatus. The two curves which might be obtained are shown plotted in Fig. 5-4. Since we cannot measure the lateral effective stresses at the completion of consolidation under each increment of vertical loading in a

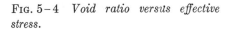

FIG. 5–4   *Void ratio versus effective stress.*

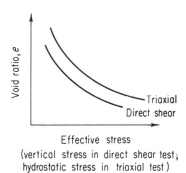

Effective stress
(vertical stress in direct shear test;
hydrostatic stress in triaxial test)

direct shear apparatus, we are unable to compute the hydrostatic effective pressure in such a test.

We shall now examine shearing test results in more detail. Suppose that in the direct-shear apparatus of Fig. 5–2 a clay soil sample first of all completely consolidates under a vertical effective stress and that then a shearing stress is applied and increased to reach a maximum equal to the strength of the soil within a few minutes after initiation. Whatever its nature within the shearing apparatus, the soil will tend to undergo either a positive or negative volume change. If it is saturated, this volume change can take place only if the pore water is free to move into or out of it during the duration of the test. We saw in Chapter 4 that, in a consolidation test on clay, 24 hr was usually allotted to each pressure increment in order to ensure relatively complete pore-water pressure dissipation. A clay soil sample in the direct-shear apparatus of Fig. 5–2 has approximately the same thickness as in the consolidation equipment, and thus no volume change can occur in a shearing test of a few minutes duration because the pore water is unable to move out of or into the sample within that period. Consequently, a positive or negative pore pressure will develop in the sample depending upon the shearing stress applied [according to Eq. (3–60)] and the properties of the material. According to the effective-stress principle of Eq. (3–52), the effective stresses developed in the soil during the consolidation portion of the test will be changed as a result of the pore pressures arising during shear, and we shall see below how this can be taken into account in the estimation of the shearing strength of the soil.

Were the soil tested a sand, drainage would be possible during shear, and therefore the applied effective stress would be maintained throughout shearing. For finer-grained soils intermediate between sands and clays some amount of drainage might take place during shear. In that case, the pore pressure (unless it were measured) and consequently the effective stress on the shearing plane would be unknown, and therefore the conditions at failure would be indeterminate. We see that the degree to which pore pressures generated in the test are dissipated depends upon the coefficient of consolidation of the soil, the thickness of the soil sample, and the rate at which the stresses are applied. The "strength" measured must be interpreted correctly in the light of these considerations. Particular caution must be taken in applying laboratory test results to field conditions because, in the field, large masses of soil are stressed, so that long drainage paths are involved, and shearing stresses, on the other hand, are quite often applied relatively slowly.

Turning to the triaxial test of Fig. 5–3, similar considerations apply when an axial stress causes shearing of the sample. Depending upon the rate of stress application, the soil properties, and the proportions and drainage provisions of the sample, the pore-water pressures generated may undergo any amount of dissipation from none to complete. Because of the size of the sample, axial stresses must be applied quite slowly to fine-grained soil samples if complete drainage is to occur during shear. The test becomes quite time-consuming, lasting, as it may, from hours to days. For these reasons it has become common laboratory practice to measure the pore pressure during a laboratory shearing test in order to estimate the effective stresses at any stage of the test. A reason why the direct-shear test tended to be displaced by the triaxial test can be seen here. The triaxial sample is completely confined, and drainage can therefore be controlled or pore-water pressure measured with relative ease, whereas in the direct-shear test, as the sample shears, additional drainage surfaces are exposed owing to displacement of the two halves of the box. Drainage conditions in the shear box are therefore difficult to control. Different types of direct-shearing apparatus have been devised to overcome this difficulty.

Two types of triaxial tests are usually employed in practice to determine the properties of saturated soils. In the first, the soil is consolidated and sheared under conditions of essentially complete drainage, so that the rate of axial stress application must be slow. The test is therefore called a *slow test* or a *drained test*. In the second, after consolidation, fairly rapid shearing is carried out without any drainage and with or without pore pressure measurement depending on the use of the results and the availability of pore-pressure-measuring apparatus. This test is called a *consolidated-quick* or *consolidated-undrained test*. A special form of the triaxial test is the so-called *unconfined compression test*, in which a cylinder of cohesive soil is compressed axially without any lateral stress being applied. No consolidation or drainage occurs before or during this test.

The time involved in testing a given soil to failure in a completely drained test depends on the coefficient of consolidation, which can be obtained from a consolidation test or from a preliminary hydrostatic pressure consolidation test on the same soil in the triaxial apparatus.

There are tables from which the rate of stress application in such a test can be calculated for soils of differing properties.

We can use Eq. (3–60) to study the pore pressure conditions in an undrained triaxial test, and the test itself can be used to give estimates of the parameters in that equation for a given soil. It will be borne in mind, of course, that the equation represents an approximately linearized representation of the behavior of any real soil, and so we must be careful to specify what type of test we are performing and to what stage of the test we are referring the calculations. We set down the equation again for convenience:

$$\Delta p_w = \frac{1}{n K_s / K_v + 1} \left(\Delta\sigma_h + \frac{K_s}{D} \Delta\tau_{\text{ave}} \right) \tag{3–60}$$

Normally, a triaxial test is performed by applying an all-around hydrostatic stress p to the sample initially, as in Fig. 5–3b, and permitting the soil sample to consolidate or drain completely under this stress increment. After consolidation, drainage of the sample is prevented by closing the drain lines (since a permeable sample may tend to drain even in a test of short duration), and additional stresses are applied, in general, to the sample. The additional stresses consist of a hydrostatic increment $\Delta\sigma_3$ followed by a stress $\Delta\sigma_1 - \Delta\sigma_3$ applied axially to the sample by the loading piston. Thus, as seen in Fig. 5–3b, the total external stresses applied to the sample at some stage in the test consist of a lateral pressure $p + \Delta\sigma_3$ and a vertical pressure $p + \Delta\sigma_1$. However, drainage was permitted under the stress component p, and therefore a pore pressure arises only from the stress increments $\Delta\sigma_3$ and $\Delta\sigma_1 - \Delta\sigma_3$.

In Eq. (3–60) applied to the triaxial test, $\Delta\sigma_h$ is the increment of hydrostatic stress applied to the soil during the stage of the test in which drainage is not permitted, and therefore in the present case it is given by the equation

$$\Delta\sigma_h = \frac{\Delta\sigma_1 + 2\,\Delta\sigma_3}{3} \tag{5–1}$$

since the two minor principal stresses are both equal to $\Delta\sigma_3$. Equation (5–1) may be rewritten in the form

$$\Delta\sigma_h = \frac{\Delta\sigma_1 - \Delta\sigma_3}{3} + \Delta\sigma_3 \tag{5–2}$$

We still have to define the shearing stress $\Delta\tau_{\text{ave}}$ in Eq. (3–60). When a soil is subjected to the general principal stress system σ_1, σ_2, σ_3, a convenient assumption is to take the average shearing stress in the soil sample to be proportional to the root mean square of the principal stress differences, as given by

$$\tau_{\text{ave}} = \tfrac{1}{3} \sqrt{(\sigma_1 - \sigma_2)^2 + (\sigma_2 - \sigma_3)^2 + (\sigma_3 - \sigma_1)^2} \tag{5–3}$$

Substituting the stresses $\Delta\sigma_1$ and $\Delta\sigma_2 = \Delta\sigma_3$ in Eq. (5–3), we find, in the present case, that

$$\Delta\tau_{\text{ave}} = \frac{\sqrt{2}}{3}\,(\Delta\sigma_1 - \Delta\sigma_3) \tag{5-4}$$

Using Eqs. (5–1) and (5–4) in Eq. (3–60), we calculate the pore pressure increment Δp_w generated by the application of the external stress increments without drainage in the triaxial apparatus:

$$\Delta p_w = \frac{1}{nK_s/K_v + 1}\left[\Delta\sigma_3 + \tfrac{1}{3}\left(1 + \frac{\sqrt{2}\,K_s}{D}\right)(\Delta\sigma_1 - \Delta\sigma_3)\right] \tag{5-5}$$

Equation (5–5) is frequently written in the form

$$\Delta p_w = B[\Delta\sigma_3 + A(\Delta\sigma_1 - \Delta\sigma_3)] \tag{5-6}$$

where

$$B = \frac{1}{nK_s/K_v + 1} \tag{5-7}$$

and

$$A = \frac{1}{3}\left(1 + \frac{\sqrt{2}\,K_s}{D}\right) \tag{5-8}$$

The substitution of the so-called *pore pressure coefficients* A and B for the terms on the right-hand sides of Eqs. (5–7) and (5–8) is made because in practice the moduli K_s and K_v in the linear form do not describe the nonlinear behavior of soil, so that A and B cannot be calculated from Eqs. (5–7) and (5–8) for a given soil. These coefficients A and B, which were first introduced by Skempton, must be obtained from pore pressure measurements in a triaxial test. It should be carefully noted that, in the reduction of the Eq. (3–60) to the abbreviated form of Eq. (5–6), the special condition of axial symmetry relating to the triaxial apparatus was taken into account, and therefore the coefficients A and B do not describe the soil alone. For this reason, the coefficients obtained cannot be employed to describe pore pressures developed under conditions in the field (for example, under long footings) in which axial symmetry is not preserved. Even when axial symmetry is preserved, as in the axial *extension* test in the triaxial apparatus, the coefficients A and B may not be used.

It was pointed out in Chapter 3 that B has the value unity when the soil is completely saturated and zero when the soil contains only air in the void spaces. The variation of B with degree of saturation is shown in Fig. 5–5 for one soil. It would be expected that, although this behavior would be simulated by other materials, different values of B would be obtained for the same degree of saturation.

The value of A can be determined by running an axial compression test on a saturated, undrained soil sample and measuring the pore pres-

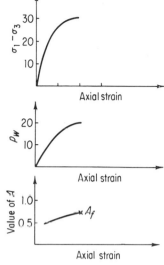

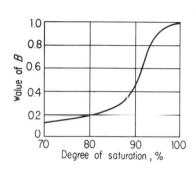

Fɪɢ. 5‒5 *Pore pressure coefficient B versus degree of saturation.*

Fɪɢ. 5‒6 *Undrained triaxial test with pore pressure measurement.*

sure generated. A typical test result is shown in Fig. 5‒6. By substituting the known stress conditions and the measured pore pressure in Eq. (5‒6), A can be calculated, and it is shown in the bottom diagram of Fig. 5‒6. It is seen that, because of the nonlinear behavior of soil, the value of A is not constant throughout the test. Indeed, investigations have shown that the pore pressure may more logically be related to the axial strain in the sample than to the axial stress difference.

Because of the variation of the pore pressure coefficient A throughout a test, it is necessary to specify the state of strain, or the stage, in a given test at which A is to be given if the test is to be used for any practical purpose. In this respect, the value of A at failure, A_f, is frequently referred to. For a saturated soil, knowledge of the value of A_f for the material obtained from a triaxial test enables the pore pressure in the ground to be estimated in situations when the stress conditions approximately correspond to those of axial compression, as, for example, on the axis of a circular or square footing.

It has been found that the value of A_f for any given soil does not vary with the magnitude of the stress conditions when the soil is in a normally consolidated state, but that A_f depends on the extent to which a soil has been overconsolidated. If we define an *overconsolidation ratio* (OCR) for a given soil to be the ratio of the highest effective stress to which the soil has been subjected to the effective stress in the soil at the time it is sheared, we find that many soils exhibit a relationship of the nature of

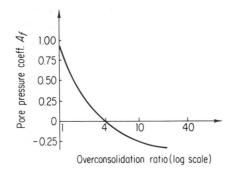

FIG. 5–7   *Effect of overconsolidation on failure value of pore pressure coefficient A_f.*

Fig. 5–7 for A_f versus the OCR. Table 5–1 gives typical values of A_f for a variety of soils.

We are now in a position to study the effect of various conditions on the shearing strength of different soils, and the next section will be devoted to that subject.

5-3   SHEARING STRENGTH FROM TEST RESULTS

The shearing strengths of cohesionless and cohesive soils differ in sufficiently many respects to make it appropriate to discuss them separately. The behavior of cohesionless materials will be discussed first.[1]

Cohesionless Soil.   When a predominantly coarse-grained soil such as sand is placed in the direct-shear apparatus at a specified void ratio and is sheared under differing vertical effective stresses, its shearing stress versus lateral-displacement behavior can be represented by curves of the type shown in Fig. 5–1. The variation from Fig. 5–1a to b depends upon the initial state of the material, as will be discussed below.

If the maximum shearing stress which can be applied to the soil is plotted versus the normal effective stress on the shearing plane for each of several tests performed on the same soil at (as closely as possible) the

TABLE 5-1   Pore-Pressure Coefficient A (at Failure; Axial Compression) for Different Soils

Type of soil	Volume change on shear	A_f
Normally consolidated clay	Contraction	$+\frac{1}{2}$ to $+1$
Compacted sandy clay	Slight contraction	$+\frac{1}{4}$ to $+\frac{3}{4}$
Lightly overconsolidated clay	None	0 to $+\frac{1}{2}$
Heavily overconsolidated clay	Expansion	$-\frac{1}{2}$ to 0

same initial void ratio and different normal stresses as shown in Fig. 5–8, it is found that the points lie close to a straight line passing through the origin of the diagram. For a different initial void ratio, a representative line of different slope appears. The material therefore exhibits essentially Coulombic frictional behavior; the *angle of friction* for each initial void ratio is given by the slope of each of the lines. Typically these slopes will vary from as low as 25° for a soil in a loose state to 45° for the same soil in a dense state. This angle of friction is calculated from the following Coulomb equation by using the observed values of peak shearing strength τ_f and effective normal stress σ':

$$\tau_f = \sigma' \tan \phi \qquad (5\text{--}9a)$$

However, in this definition of ϕ, as we shall see below, the angle contains both frictional and volume change components.

For one granular soil, it is found that the curves of shearing stress versus deformation such as shown in Fig. 5–1 for different initial states tend toward the same *ultimate* value of shearing stress. This is the stress at which no further change in volume of the material is observed to occur. For the normal effective stress applied, the void ratio of the material at this stage is called the *critical void ratio*. Thus, in the course of shearing, the loose soil becomes more dense to reach, ultimately, the critical void ratio, whereas the dense soil expands to the critical value. The critical void ratio depends on the applied effective stress. We can now more objectively define "loose" and "dense" states of a soil with reference to void ratios above or below the critical value at a given stress level.

Because of the expansion or contraction of the material on shear, the horizontally applied shearing force has components tangential and normal to the surface determined by the ratio of the vertical motion of the soil sample to the horizontal motion at any stage of the process, $\tan \alpha$; this is illustrated in Fig. 5–9. Thus, as pointed out earlier, a portion of the shearing stress required to cause movement of the top half of the box goes into doing work against the vertical stress in a test on a dense soil. By resolving forces normal and tangential to the plane of sliding and by

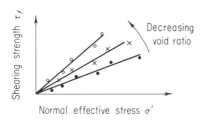

FIG. 5–8 *Shearing strength of cohesionless soil.*

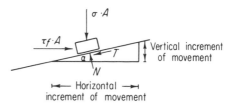

FIG. 5–9 *Effect of expansion on work done in direct-shear test.*

using the Coulomb frictional equation $T = N \tan \phi_f$, where ϕ_f is a "true" angle of friction of the soil, we can obtain an equation as follows, separating the shearing stress into components (1) required to overcome friction in the material and (2) necessary to do dilatational work (the latter may be positive or negative, depending on the initial state of the soil).

$$\tau_f = \sigma' \tan (\phi_f + \alpha) \tag{5-9b}$$

The angle of internal friction of the soil ϕ of Eq. (5–9a), as measured in the test, can then be seen to result from the two components ϕ_f and α as in the equation

$$\phi = \phi_f + \alpha \tag{5-10}$$

Given a direct-shear test of the type discussed above whose results are plotted in the manner of Fig. 5–1, it is possible to separate these two components when the volume increase or decrease of the specimen has been measured. When this is done, it is found that the "true" friction angle ϕ_f of the material is relatively constant and independent of its loose or dense state. An expression similar to Eq. (5–9b) can be derived for the triaxial test but will not be discussed here. Subtraction of the volume change component from the angle of friction as usually computed is occasionally termed the *boundary energy correction*.

In practice, the friction angle obtained for a given soil from Fig. 5–8 represents the actual stress required to shear a granular soil at a given normal stress, since both shearing and volume change work must be done.

It is instructive to plot the stress variations during both a triaxial and a direct-shear test on sand on the Mohr diagram of Chapter 3 in order to compare the results obtained. At the beginning of the test in a direct-shear machine the vertical stress σ is the major principal stress and may be plotted in the Mohr diagram of Fig. 5–10a. However, the two lateral principal stresses, presumably equal for a symmetrical box, are not known, and consequently the initial Mohr circle cannot be plotted. As known

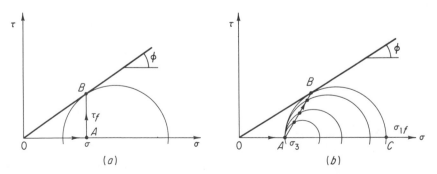

Fig. 5–10   *Comparison of direct-shear and triaxial test results. (a) Direct-shear test. (b) Triaxial test.*

shearing stresses are applied to the soil sample, the normal stress on the midplane (point A in Fig. 5–10a) of the box may be assumed to remain constant but is no longer a principal stress, so that we can draw in the Mohr space of Fig. 5–10a a *stress path OAB* showing the changing stress conditions only *on the midplane* of the direct-shear test. Other principal stresses or directions and the stress conditions on other planes in the box are unknown, and therefore the circle of stress on the Mohr diagram cannot be plotted.

At failure, however, the stress path ends at the point B at which the shearing stress has reached its failure value τ_f. Since, on this midplane, the behavior of the material follows the Coulomb law of friction, the stress obliquity will have reached a maximum at point B in Fig. 5–10a. From Chapter 3, the line joining point B to the origin represents the resultant stress on the failure plane at failure, and the angle ϕ this plane makes with the σ axis is the obliquity of stress on the failure plane. On no other plane in the test specimen can the obliquity exceed the angle ϕ. Consequently, we may reason that the Mohr stress circle in Fig. 5–10a describing the stress conditions at failure must pass through point B, have its center on the σ axis, and must be tangential to the line OB at B. This circle is shown in the figure. The Mohr stress circles for intermediate stress states cannot be deduced. Since the line OB is tangential to all circles at failure for one initial void ratio, it is frequently called the *failure envelope*.

The above conditions may be contrasted with those obtained in the triaxial test, which are shown in Fig. 5–10b. Let us carry out a completely drained test. Initially a hydrostatic stress σ_3 is applied to the soil sample which is permitted to drain under this stress. To this stage the stress path in the Mohr diagram of Fig. 5–10b therefore lies along the σ axis to point A. Now an axial stress is applied, again with drainage, to develop the major principal stress σ_1. The minor principal stress remains as indicated by point A and, as σ_1 increases, successive circles, anchored at A, can be drawn to represent the soil stress states. Failure occurs at a particular value of major principal stress σ_1. Unlike the direct-shear test, the stress conditions are known for each stage of the test.

If we carry out more tests at different values of σ_3 on the same soil at the same initial void ratio, we shall find in each case that failure occurs when the stress circle becomes tangent to a straight line through the origin, the failure envelope OB in Fig. 5–10b. The point of tangency B in a particular test represents the stress conditions on the plane in the soil sample on which the obliquity is a maximum; this is called the *failure plane*. Since point A is the pole of the diagram, the orientation of the failure plane is given by the line AB from the Mohr circles developing in a test; the stress condition on the future failure plane can be followed. These stress conditions are represented by points along the path AB in

Fig. 5–10*b*. It is seen that the triaxial test is much more informative regarding stress conditions than the direct-shear test.

It is also interesting to analyze the conditions in an *undrained* triaxial test when the sample is first stressed hydrostatically under a stress σ_3 with drainage permitted and is then subjected to shearing stresses without drainage. If we measure pore pressures in a test on a soil in a relatively loose condition, we shall find, according to previous considerations, that the pore pressures increase throughout the test. Since it is a hydrostatic pressure, we can represent the pore pressure at any stage of the test by a distance OO' along the σ axis from point O in the Mohr stress space. In effect, the point O represents the origin from which total stresses are measured, whereas, during a test, O' can be considered a transient origin to which effective stresses are referred.

Initially, in the present test, when the total stress σ_3 is applied to the sample with drainage, the stress condition in the soil is represented by point A in Fig. 5–11*a*; at this time the distance OA represents both total and effective stresses in the soil sample, since the pore pressure is zero. At a later stage when the value of the major principal total stress is σ_1 and is given by OB on the figure, a pore pressure p_w has been developed as a result of shearing, and this is plotted as the distance OO' along the σ axis. The Mohr stress circle, whose diameter, being proportional to the principal stress difference, does not depend on the pore pressure, is shown in Fig. 5–11*a*.

At failure, the pore-water pressure has reached the positive value p_{wf} and an appropriate distance OO'' is laid out along the σ axis. At this state, the major principal stress has reached the value σ_{1f} and is shown by point C; the Mohr circle through points A and C is also shown. Thus, at failure the total principal stresses are OA and OC, whereas the effective stresses are $O''A$ and $O''C$.

If the test was carried out with no knowledge of the pore pressures

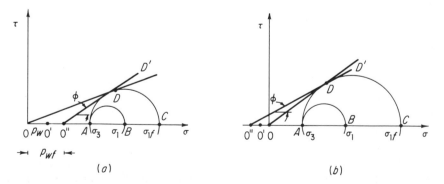

Fig. 5–11   *Behavior of cohesionless soil in undrained triaxial test.* (a) *Loose cohesionless soil.* (b) *Dense cohesionless soil.*

generated, only total stresses could be recorded, and thus the only failure envelope we could draw would be the line *OD* through the origin *O*. The angle which such a line makes with the σ axis is called the *apparent angle of internal friction* for the soil. In actual fact, the effective stresses are the important ones, and therefore the friction angle of the soil is truly represented by the line $O''D'$ in the figure. It is seen that the real angle of internal friction in a loose cohesionless material is greater than the apparent angle.

When a dense cohesionless material is tested, the conditions during stressing in the triaxial apparatus are as represented in Fig. 5–11b. The reasoning follows that associated with Fig. 5–11a, except that the shearing of a dense soil (which is trying to expand) tends to cause the development of negative pore pressures which, with the sign convention of the Mohr diagram, are plotted to the *left* of the origin along the σ axis in Fig. 5–11b. At failure it is found that the apparent angle of internal friction of the dense material, given by *OD*, is very much larger than the true angle given by $O''D'$ in Fig. 5–11b. Thus, in shearing, a dense soil has a high strength because of the increased effective stresses developed by the negative pore pressures, whereas a loose cohesionless soil is much weaker as a result of the positive pore pressures generated.

In practice, both of these tests may be considered to represent the stress conditions in situations where the rapid shearing of granular material occurs when drainage is inhibited by the combination of the following factors: rate of shearing, soil permeability, and the length of drainage path available. It has been observed in the case of loose cohesionless materials that rapid shearing may result in the loss of all effective contact between grains so that the soil flows as a dense but relatively inviscid liquid. This occurrence, called *soil liquefaction*, may occur in loose saturated sands on the application of sudden stresses due to earthquakes, explosions, or other dynamic phenomena. It even takes place in a dry loose sand or silt when stresses are applied very rapidly.

In this context, it may be repeated that the soil is described as dense or loose with respect to the critical void ratio of the material, and this depends on the hydrostatic effective stress. Consequently, if a homogeneous sand layer should exist in nature, with a void ratio constant with depth, the increase of hydrostatic pressure in the soil with depth (due to the overburden) makes it possible for the soil to be denser than the critical void ratio down to a certain depth, below which it is looser than critical. The critical void ratio can be approximately determined from a shearing test in a triaxial apparatus.

Attempts have been made to predict theoretically the shearing and deformational behavior of sands based upon analyses of assemblages of spherical grains of identical size, but they have generally been limited by the complexity of the mathematics of the problem. Some generalizations, however, have been made from these analyses, and research work is still being done on this approach. For practical purposes, the concept of an angle of internal friction has been found to be of value and is applied to all conditions of soil stressing. The representation of failure by a straight line on the Mohr diagram is sometimes called the Mohr-Coulomb failure hypothesis.

If a cohesionless soil sample is subjected to a hydrostatic stress in the triaxial apparatus, with drainage permitted, and then, before shearing, the hydrostatic stress is reduced, again with drainage, it is found that the failure Mohr circle and envelope obtained on shearing is essentially the same as that developed for a test carried out after stressing to the lower hydrostatic stress only. Thus it can be approximately said that the failure behavior of the soil does not depend on the stress "history" of the material. The behavior of cohesive soils is different and will be described below.[2]

Fine-grained Soils; Cohesion.   If a very fine-grained saturated soil containing clay sizes is normally consolidated under a hydrostatic stress σ_3 in the triaxial test and is then sheared by increasing the axial stress with drainage, a failure Mohr circle is obtained. A series of such tests, all performed on normally consolidated soil at differing hydrostatic effective stresses, gives rise to a number of circles on the Mohr diagram as shown in Fig. 5–12a. All of the circles are found to be tangential to a straight line in the way exhibited by the cohesionless material. It would appear, therefore, that the soil demonstrates frictional behavior identical with that of a cohesionless soil, and, in fact, the angle of the straight line of Fig. 5–12a with respect to the σ' axis is called the *angle of internal friction* of the clay. Such tests do not, therefore, demonstrate that the material possesses cohesion. This, however, is not the only possible interpretation of the material's behavior.

We saw in Chapter 1 that the microscopic particles of clay form arrangements which depend greatly upon the nature of the particles and

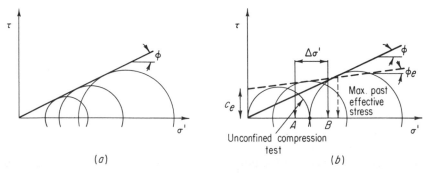

Fig. 5 –12   *Behavior of clay soil in drained triaxial test. (a) Normally consolidated clay. (b) Overconsolidated clay.*

their electrolytic environment. In most natural clays it was considered that the particles of clay touch each other to form a complex structure. The forces of interaction between the particles are important in this respect. Consequently, when a sample of clay is gradually subjected to an increase in applied stress and is permitted to drain, the clay platelets bend, slip, and translate with respect to each other until a structural framework is formed which is relatively stable (time-dependent effects are ignored here) under the increased effective stress.

Therefore we have a soil in which the number of contacts per unit volume of the soil depends on the effective stress applied to the soil. Since we may expect that the shearing strength of such a material is related to the number of particles per unit volume, we may deduce that the shearing strength of the material will be increased as a result of a larger number of contacts. Thus, we can see that an increase in applied effective stress gives rise to an increased number of contacts which, in turn, give the material an increased resistance to deformation and therefore a higher shearing strength. It does not follow, therefore, that the sloping straight line of Fig. 5–12a necessarily represents a simple frictional behavior of a clay mass with respect to effective stresses in the same way as for sands. Further, because of the nature of the forces and interactions between particles, many of the contacts which are made in this way are irreversible. In other words, when the applied stress is removed, again with drainage, the contacts developed at the higher stresses do not all separate nor does the material return to its original structure.

This point may be illustrated in more detail by considering the observed behavior of overconsolidated clays. Let us take the clay specimen whose normally consolidated behavior is represented by Fig. 5–12a and subject each of a number of samples of that clay to a relatively high hydrostatic stress with drainage. Next, we unload the samples to different hydrostatic stresses and permit each one to swell (drain) to its equilibrium state at the lower pressure. As we noted in Chapter 4, the unloading void ratio versus pressure curve is different from the normal consolidation curve. The void ratio at a given hydrostatic effective stress then depends greatly on the highest hydrostatic pressure to which the soil has been subjected as a result of contact and structure irreversibility.

Now each sample of clay is sheared in a triaxial test under drained conditions. The results of the tests are plotted in Fig. 5–12b, and it is found that the strength of the soil as indicated by the envelope of the Mohr circles is not proportional to the existing hydrostatic effective stress at the start of the test. The envelope, in fact, is not found to be a straight line. Even in a test in which no lateral stress whatsoever is applied, an unconfined compression test, an axial stress is still required to fail the soil. We say of this material, which exhibits shearing strength in the apparent absence of an external hydrostatic stress, that it possesses cohesion. We have indicated that the strength of the material is related to the number of contacts developed between particles under the highest effective stress. In addition, most of the contacts produced by an effective stress remain after the stress is removed. We can explain the strength of the overconsolidated soil found in the test sequence above by this argument. It appears that the cohesiveness of the sample is related both to the maximum effective stress to which the soil has been subjected in the past and to the effective stresses at the time it is sheared to failure. The strength of the soil thus depends on its stress history in contrast to the behavior of coarse-grained soils.

In practice, we are interested in the shearing strength of an overconsolidated soil when we apply an additional stress increment to the soil as a result of a construction process. Only drained conditions are considered at present. Consequently, we may be concerned, in a practical situation, with a range of effective stresses from that existing in the soil at point A, for example, in Fig. 5–12b to that at point B in the same figure, in which the distance AB represents the effective-stress increment resulting from our construction process. For complete drainage of the material under the applied stresses, it is convenient to represent the strength envelope by means of a straight line best fitting the actual curved Mohr envelope between the values of stress represented by points A and B. This straight line has a slope tan φ_e and, when produced, an intercept on the τ axis of c_e. The angle φ_e is again termed the angle of internal friction of the soil (with respect to effective stresses). The intercept c_e is frequently referred to as the "effective cohesion" of the soil, although we can see that it has little direct relation to any real property of the material, being merely a geometrical consequence of the choice of a best-fit straight line. For the normally consolidated material the angle of internal friction with respect to effective stresses will be different from this value, as seen in Fig. 5–12b.

Now we must proceed to consider the observed failure behavior of cohesive soils when subjected to undrained shear tests. In the triaxial test a sample of clay is normally consolidated under the hydrostatic effective stress σ_3, and following this an axial stress is applied without further drainage. Since the permeability of the fine-grained soils is small, drainage is effectively restricted by applying the axial stress rapidly enough. (Loading times lasting from several minutes to tens of minutes are "rapid" in this sense.) Previous considerations have indicated that a positive pore pressure is generated in such a test, so that the behavior of the material parallels that of the loose sand, and at failure we obtain a stress circle as shown in Fig. 5–13a. If we measured the pore pressure at failure p_{wf}, we could locate the origin O' for effective stresses at failure.

If another sample of identical soil is also first consolidated to the stress σ_3 and then, *without drainage*, is subjected to an increased hydrostatic total pressure before shearing, we shall find that the Mohr circle at failure

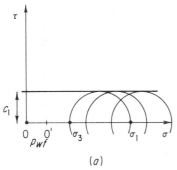

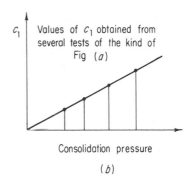

(a) (b)

FIG. 5–13 *Behavior of normally consolidated clay in undrained triaxial test.*
(a) Mohr circles at one consolidation pressure. (b) Apparent cohesion versus con-
solidation pressure.

(for a *saturated* soil) has the same diameter as in the first case, since no
change in effective stress has been permitted and the stress increase has
been taken by the pore water. The series of circles from this consolidated-
undrained test have a horizontal envelope on the Mohr diagram of total
stresses, and the envelope has an intercept c_1 on the τ axis. The value c_1,
called the *apparent cohesion* of the soil, obviously depends on the original
consolidation pressure. When tests are carried out on the soil consolidated
to different pressures, the resulting intercepts may be plotted on a diagram of apparent cohesion c_1 versus the consolidation stress, as illustrated
in Fig. 5–13b. In practice, this corresponds to the behavior of a soil which
has been consolidated to differing stresses in the ground (as a function of
depth) and is then subjected, as a result of construction, to increase of
both hydrostatic and shearing stresses rapidly enough that no drainage
can take place. The rate of applied-stress increase can in reality be very
low to meet this requirement because of low permeabilities and long
drainage paths. The profile of strength with depth in the ground can be
shown by a diagram similar to Fig. 5–13b in which the horizontal axis
represents the effective stress increasing with depth.

Each of the consolidation pressures in figures such as Fig. 5–13a represents a soil sample in a normally consolidated state. This diagram is
therefore qualitatively equivalent to Fig. 5–12a. A soil, however, may
also exist in an overconsolidated state before it is sheared without drainage, and this situation is represented by Fig. 5–14, in which total normal
stresses are plotted on the horizontal axis. To obtain this figure a soil
sample is placed in a triaxial apparatus and is then normally consolidated
to some effective stress σ'_c and sheared without drainage. This gives circle
1 in Fig. 5–14. To obtain the other circles, samples are first consolidated
under the effective stress σ'_c and are then permitted to swell at a lower
stress to an overconsolidated state. When swelling is complete, drainage

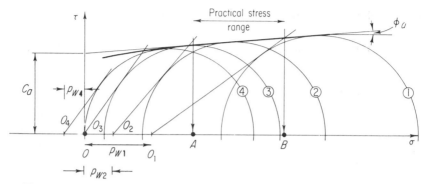

FIG. 5-14   *Behavior of overconsolidated clay in undrained triaxial test.*

is prevented and the soil is sheared once again to give circles 2 to 4, obtained for different rebound pressures.

The circles may be explained as follows. At failure, under the stress conditions of circle 1, a positive pore pressure p_{w1} has been generated, the origin of effective stresses is O_1, and consequently the effective stresses in the soil at failure are smaller than the total stresses. When the sample has been overconsolidated to a small extent (refer to circle 2), a somewhat smaller, but still positive, pore pressure is generated; it is represented by p_{w2} in the diagram. The effective stresses for the failure circle 2 are measured from the origin O_2. With increasing degree of overconsolidation, the soil tends to contract less and less on shearing, so that circle 3 may represent a state of overconsolidation at which no pore-water pressures at all are generated at failure. It will therefore be seen that, on an effective-stress basis, circles 2 and 3 represent material only very slightly weaker than that described at failure by circle 1. When the overconsolidation ratio is high, the sample will tend to develop negative pore pressures during shear; this is represented in the figure by p_{w4}, and the origin of effective stresses becomes point O_4 on the diagram. Consequently, all of the tests carried out in this manner represent virtually equivalent effective-stress conditions for the material, and therefore the envelope to the circles is almost horizontal as shown.

When the soil is known to exist in an overconsolidated state in nature, the addition to the total stress state which will be caused by construction in a time so short that drainage cannot occur can be represented in Fig. 5-14 by a range of stresses (as in Fig. 5-12*b*) from point A to point B. Once again the failure envelope over this stress increment can be represented by a straight line with a τ axis intercept c_a and slope $\tan \varphi_a$. In this case φ_a is referred to as the apparent angle of internal friction and the intercept c_a is termed the apparent cohesion of the material.

It can be seen from the above results that the shearing strength of soils in general can be represented by the equation

$$\tau = c + \sigma \tan \varphi \qquad (5\text{-}11)$$

in which, for a cohesionless soil, $c \cdot = 0$ and, for a cohesive soil, c has values which differ depending upon the stress conditions and the practical application involved. The stress σ may either be the total or the effective stress, again depending upon the considerations of the test and its application. The values of c and φ are chosen appropriate to the initial state of consolidation of the soil and the drainage conditions under which it will be sheared. Equation (5-11) is usually applied to a given surface in the soil on which failure is thought likely. The stress σ is the stress normal to that surface, and τ is then the maximum shearing stress which can be sustained.

5-4 EFFECT OF RATE OF SHEARING ON STRENGTH

Because of dynamic stresses generated in soil as a result of machinery, earthquakes, and nuclear and other explosions, it is becoming necessary to study soil behavior at increased rates of stress application. In the present context we are interested in studying the effect of shearing a sand or clay soil at various rates, but it has been found that the problem is a very difficult and complicated one. A great deal of research is currently being directed toward an understanding of the behavior of soil under rapid loading. The mechanisms in the soil are complicated because of pore pressure and acceleration effects. However, it appears that the true shearing strength of a cohesionless soil is almost independent of the rate at which the soil is tested, whereas the shearing strength of a cohesive soil increases somewhat as the rate of application of the stress increases.

On the other hand, the behavior of soils under shearing stresses which are less than the short-term or static soil strength but are applied for long times is also of interest from the point of view of the long-term safety of structures or the stability of slopes over periods of many years. Because of viscous creep, clay will deform continuously for a long time under the application of a shearing stress much smaller than the value which will cause failure in a test carried out at normal laboratory rates of strain. When the strain progressively increases with time in this way, the deformations obtained may ultimately be harmful to structures supported on the material or the soil material may itself ultimately fail. An example of this is shown in Fig. 5-15, which represents various stress levels applied in drained tests to one type of clay soil for different times. It is seen that even a relatively small stress will cause the material to fail after several days or months have passed. It would be logical to extrapolate the clay

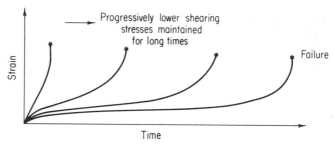

FIG. 5–15   *Long-time strain behavior of clay.*

behavior obtained in such tests to much longer times (even to geological times) in consideration of the long-term stability of slopes.

PROBLEMS

5-1   Explain why the two curves of Fig. 5–4 are different.

5-2   What are the relative advantages and disadvantages of (a) the direct-shear test and (b) the triaxial test?

5-3   If a sample of soil with a unit weight of 100 pcf has been taken for strength testing from 20 ft below the ground surface, what stress conditions would you apply to it in (a) the direct-shear apparatus and (b) the triaxial equipment before shearing it to simulate the conditions in the ground? Water table is at ground surface.

5-4   If the sample of soil in Prob. 5–3 comes from a thin clay layer and you are testing its strength in order to design a highway fill which will be built in a short time on the ground surface, what kind of shearing test would you carry out?

5-5   Alternatively, if the soil of Prob. 5–4 is to be stressed by an earth dam which will be constructed over a period of 3 years, what kind of shearing test would you carry out in this case?

5-6   Calculate the pore pressure which would be developed in a normally consolidated, saturated clay in a consolidated-undrained test with a constant σ_3 pressure of 10 psi and in which the total axial stress reached a value of 35 psi at failure of the sample (see Fig. 5–7).

Answer   15 to 20 psi

5-7   What value would the pore pressure reach at failure (at the same stresses) if the soil was 90 percent instead of fully saturated?

Answer   6 to 8 psi

5-8   At what overconsolidation ratio do you expect to find no pore pressure generated at failure in an undrained test on a clay soil?

5-9   In a test on the soil of Prob. 5–8 do you expect pore pressures to be generated as the axial stress is increased to failure?

5-10   Resolve the forces in Fig. 5–9 into their components normal and tangential to the plane of sliding and establish that Eq. (5–9) is correct.

5-11   In a direct-shear test on a medium-dense sand, a shearing stress of

8.5 psi is recorded at failure. If the friction angle of the soil is 38°, what was the normal stress on the failure plane? *Answer* 10.9 psi

5-12 What were the values of the major and minor principal stresses and the maximum shearing stress in the test of Prob. 5–11? *Answer* 28.4; 6.8; 10.8 psi

5-13 In the test of Prob. 5–11 it was observed that the upper half of the shear box had risen 0.05 in. at failure because of the dilatancy of the sample. The relative horizontal displacement of the two halves of the box at failure was 0.5 in. What was the true angle of friction of the soil? *Answer* 32.3°

5-14 In a triaxial test on the same sand of Prob. 5–11, at the same void ratio, a value of $\sigma_3 = 10$ psi was chosen. At what *additional* axial stress would you expect failure to take place in a drained test? *Answer* 31.8 psi

5-15 In an *undrained* triaxial test on the soil of Prob. 5–11 (again at the same void ratio) a pore pressure of 3.2 psi *below* atmospheric pressure was recorded at failure (-3.2 psi) when the value of lateral stress was again 10 psi. What was the apparent angle of internal friction in this test? *Answer* 33.1°

5-16 In a drained triaxial test on a normally consolidated clay soil, a major principal stress of 47 psi was required to fail the soil when the lateral stress was 15 psi. What was the angle of internal friction of the soil? *Answer* 31.1°

5-17 If the stresses of Prob. 5–16 were required to fail a clay soil in an unconsolidated-undrained test, what was the cohesion of the material? *Answer* 16 psi

5-18 In an unconfined compression test on an overconsolidated clay an axial stress of 21 psi was required to fail the soil. In a consolidated-undrained test on the same soil at a confining pressure of 12 psi, an axial stress difference of 28 psi was required at failure. If this stress range was appropriate to the field problem you were interested in, what values would you use for apparent cohesion and apparent angle of friction for the soil? *Answer* 8.1 psi and 12.5°

5-19 If the normally consolidated angle of internal friction of the same soil as in Prob. 5–18 was 26°, what was the maximum past effective stress in Prob. 5–18? *Answer* 30.9 psi

5-20 With the result of Prob. 5–19 and using Fig. 5–7, what pore pressure do you estimate was generated at failure in the consolidated-undrained test of Prob. 5–18? *Answer* About 5.5 to 6.0 psi

REFERENCES

1. R. F. Scott, "Principles of Soil Mechanics," Addison-Wesley Publishing Company, Inc., Reading, Mass., 1963.
2. D. W. Taylor, "Fundamentals of Soil Mechanics," John Wiley & Sons, Inc., New York, 1948.

6 | Failure in Soils

In any type of construction involving soils, two types of problem involving the deformation of the soil may be met. The first of these concerns the total and relative amounts of settlement which will take place under a structure, for any reason, without failure of the soil. Consideration was given to this situation in a preceding chapter. The possibility also exists that the weight of the structure on the ground will generate stresses which exceed the yield or shearing strength of the ground. This is the second class of problem and will be discussed in this chapter.

In Chapter 3 we established a basic set of equilibrium equations (3–4) for any elemental soil mass. In the same chapter, equations representing the stress-strain behavior of a linearly elastic material were also set down and a class of problems of elastic equilibrum was set forth by substituting the elasticity relations in the equilibrium equations. In a similar way, the Mohr-Coulomb equations of Chapter 5 describing the failure conditions in a soil could also be substituted in the equilibrium equations to give expressions for the state of stress in a soil mass on the point of failure but still at equilibrium. This stress condition is referred to as the *state of plastic equilibrium*. Because of the unlimited deformations which may take place at yield, no compatibility condition has to be satisfied in problems of plastic equilibrium. However, in the classical theory of plasticity the material is assumed to yield at constant volume. As we have seen, this last condition may not be true in a failing soil, but we shall ignore it here.

When the equations of plastic equilibrium are obtained, they are found to be more complicated than the corresponding elastic equations, and as a result they are very difficult to solve under practical boundary conditions except in certain particularly simple cases. The solutions to some of these simple cases will be discussed in this chapter. In more realistic failure problems approximate solutions must be obtained not by considering the general equations of equilibrium and yield of a single element, but by considering the equilibrium of a large mass of soil and the yielding conditions along certain surfaces on which shearing failure is likely to occur. This method of solution will also be considered in this chapter.

6-1 STABILITY OF SOIL MASSES

Although yielding in soil masses occurs in many practical conditions in three dimensions, these problems are generally too difficult to analyze except very approximately. Most practical analyses of failure confine themselves to two dimensions, and in the present chapter we shall therefore consider two-dimensional failure conditions only.

It will be recalled from Chapter 3 that the two-dimensional equations of equilibrium were represented by a Mohr circle of stress, and we also saw in Chapter 5 that, under many conditions of stressing in two dimensions, failure in soils could be adequately described by a line or envelope on the Mohr diagram. Consequently, all circles tangential to the envelope describe stress states which satisfy both equilibrium and yield requirements for a soil element. Thus, instead of considering the complicated equations of plastic equilibrium, we can illustrate some of the simple cases of yield or failure in soil masses graphically by referring to the Mohr diagram in the special circumstance when the Mohr circle touches the yield envelope for the material.

Many practical cases of yielding in soil masses can be represented by specialization of the general diagram, Fig. 6-1, which shows a soil mass defined by two straight-line surfaces including the angle α. The mass is assumed to be on the point of yielding under some distribution of stresses applied to the two surfaces. In general, the stresses will act at some angle to the surface, so that both normal and shearing stresses will be involved. For simplicity here, however, we shall show normal stresses only. Along one face of the soil mass in Fig. 6-1 a stress q is applied, and a larger stress p is acting on the other face. In most soil problems, the magnitude

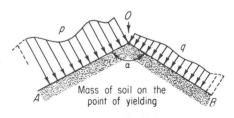

Fig. 6-1 *Limiting equilibrium in soil mass.*

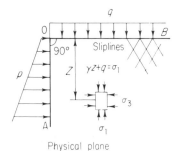

FIG. 6-2   *Special case of limiting equilibrium (active earth pressure).*

of the stress q will be known from either the structural load or the soil conditions and we desire to find a maximum value of p which will just cause the soil in the region AOB to fail.

Since yield or failure in a soil is a phenomenon involving friction, as described by the Coulomb law, there will be two values of p for which failure in the mass AOB can occur. One value of p to cause yielding will be larger than q, as shown in Fig. 6-1, and will result in displacement of the mass of soil in the general direction of the face OB, which will tend to move outward or upward under failure conditions. Another possible failure condition would develop in the soil mass were the load p reduced to a magnitude smaller than q such that the soil mass tended to yield by moving to the left in the diagram, displacing the face OA upward and to the left. These two yielding conditions are called the two states of limiting plastic or yield equilibrium, and many soil failure problems can be represented by one or the other of the two states, as we shall discuss further.

One simple special case of the general situation shown in Fig. 6-1 is given in Fig. 6-2, in which the angle α is made equal to 90° and the plane surface OA may then represent, for example, a retaining wall. The surface OB is then the surface of the ground retained by the wall, and it may be subjected to a stress or load q arising as a result of the weight of a structure or a pile of material such as ore.

The two limiting values of the stress p in this condition then represent the maximum and minimum values of p which may be applied to the retaining wall in order to cause failure of the soil mass inside AOB. Since, in many cases in practice, retaining walls located at OA in Fig. 6-2 are relatively flexible or are built on foundations which may slide outward away from the soil mass, one consideration in designing such a wall is to determine what will be the minimum soil pressure placed on the wall (i.e., the minimum value of p). This pressure will occur when the wall moves away from the soil, permitting the soil to yield. This circumstance is one of the two states of plastic equilibrium, and the pressure p at which yielding just occurs is referred to as the *active pressure*. We shall consider this situation in more detail.

6-2 ACTIVE EARTH PRESSURE

Since Fig. 6–2 illustrates the distribution of pressures acting on the soil mass whose geometrical proportions are also indicated, it is said to show the *physical plane*. The stresses can also be shown in a Mohr diagram (Fig. 6–3), which is then described as the *stress plane*.

In Fig. 6–3 the yield envelopes for the general case of a soil possessing both cohesion c and an angle of internal friction φ are shown on the Mohr diagram. At any depth z below the horizontal surface OB of Fig. 6–2 the vertical stress on a soil element will be a principal stress σ_1 equal to $\gamma z + q$ when no shearing stresses act at the surface. If we assume that the initial conditions are those existing in a semi-infinite mass of soil (i.e., the presence of the wall is taken to have no influence on the stresses), the lateral pressure is usually smaller than the vertical pressure and will, in the case when there is no friction on the surfaces, also be a principal stress σ_3.

The ratio of the lateral to the vertical pressure at any depth is called the *lateral earth pressure coefficient*. In the case that the soil has the relatively simple stress history of being deposited in successive layers one on top of the other, without subsequent removal, the earth-pressure coefficient is said to have an "at-rest" value and is given the symbol K_0. In sands this at-rest earth pressure coefficient may have a value near 0.5, whereas in clays it can be as high as unity. When the vertical stress in a soil is decreased, as a consequence of removing surface material, say by erosion, the lateral stresses decrease less, so that higher at-rest earth pressure coefficients may be obtained in these overconsolidated soils. Values as high as 2.5 have been estimated to exist in some overconsolidated clays.

Thus, when the soil element is not on the point of failure, the stress conditions at equilibrium may be represented by the circle on ab as diameter in Fig. 6–3. Here Oa represents the major principal stress σ_1 and Ob represents the minor principal stress σ_3 acting in the soil and also the lateral stress on the wall. If we permit the wall to displace to the left in such a way that the lateral stress on all soil elements decreases proportionately with depth, then b will move to the left in Fig. 6–3 while the vertical stress remains unchanged. Eventually a limiting lower value of

Fig. 6–3 *Stress plane of Fig. 6–2.*

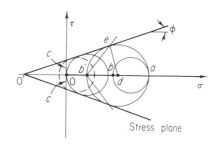

Stress plane

lateral stress will be reached when the stress circle touches the two failure envelopes of the diagram. At this stage, the circle has diameter ab' in Fig. 6–3 and represents the stress conditions at yield at the given depth in the soil. No further reduction in lateral pressure can be made, so that even if we let the wall move outward at this value of pressure, the circle ab' still continues to represent the at-yield stress conditions at the depth z. Since the vertical principal stress is proportional to depth z and the envelopes on Fig. 6–3 are also straight lines, circles representing yield conditions at other depths will be proportional in size to the depth. By drawing the radius de of the failure circle from its center d to the point of tangency e, we see that

$$\sin \phi = \frac{de}{O'd} \tag{6–1}$$

However

$$de = \frac{\sigma_{1f} - \sigma_{3f}}{2} \qquad \text{radius of failure circle}$$

when we add the subscript f to indicate that failure conditions are being considered, and

$$O'd = O'O + Od = c \cot \phi + \frac{\sigma_{1f} + \sigma_{3f}}{2}$$

Putting these relations in Eq. (6–1), we get

$$\sin \phi = \frac{(\sigma_{1f} - \sigma_{3f})/2}{(\sigma_{1f} + \sigma_{3f})/2 + c \cot \phi}$$

which may be rearranged to give

$$\sigma_{3f} = \frac{1 - \sin \phi}{1 + \sin \phi} \sigma_{1f} - \left(\frac{1 - \sin \phi}{1 + \sin \phi}\right)^{1/2} 2c \tag{6–2}$$

In Eq. (6–2) the major principal stress σ_{1f} is, as before, equal to $\gamma z + q$. If we place $c = 0$ in Eq. (6–2), we obtain the form of the equation which represents the special case of a soil possessing internal friction only, as follows:

$$\frac{\sigma_{3f}}{\sigma_{1f}} = \frac{1 - \sin \phi}{1 + \sin \phi} = K_a \tag{6–3}$$

In this case the ratio of the lateral principal stress to the vertical principal stress at failure is called the *active earth pressure coefficient* and is given the symbol K_a.

Another special case is that in which the soil possesses cohesion only (angle of internal friction equals zero) and the appropriate equation is obtained from Eq. (6–2) by putting $\phi = 0$.

$$\sigma_{1f} - \sigma_{3f} = 2c \tag{6–4}$$

In this case, and also in the general case of the soil possessing both cohesion and friction, it can be seen from Fig. 6–3 that at points near the surface the position of point a on the Mohr diagram will be such that construction of the failure circle will lead to a tensile minor principal stress. This is shown, for example, by the dashed circle in the diagram. In this case, the minimum value of Oa occurs at the surface of the soil and must be equal to the stress q. Here the minor principal stress at the ground surface has the value given by

$$\sigma_{3f} = \frac{1 - \sin \phi}{1 + \sin \phi} q - \left(\frac{1 - \sin \phi}{1 + \sin \phi}\right)^{1/2} 2c \tag{6-5}$$

so that σ_{3f} may be compressive or tensile, depending upon the values of q and the soil properties. If σ_{3f} is tensile, the state of tension will exist down to a certain depth z_0. We can obtain this depth by putting $\sigma_{3f} = 0$ and $\sigma_{1f} = \gamma z_0 + q$ in Eq. (6–2) and solving for z_0, to get

$$z_0 = \frac{2c}{\gamma}\left(\frac{1 + \sin \phi}{1 - \sin \phi}\right)^{1/2} - \frac{q}{\gamma} \tag{6-6}$$

Generally speaking, soils are not considered to be capable of resisting tensile stresses for other than very short times, so that in design calculations it is usually assumed that a crack will develop to at least the depth z_0 in Fig. 6 4, and the tensile stress in the zone to depth z_0 is ignored in calculations of the forces acting on, for example, retaining walls.

The distributions of lateral pressure with depth for the three cases (1) soil possessing cohesion and friction, (2) soil possessing no cohesion, and (3) soil with cohesion only are shown in Fig. 6–4, in which no surcharge load q is shown.

It must be clearly understood that the stress distributions which have just been discussed hold only under the special conditions described. In summary, these stresses hold when no tangential or shear stresses are applied to the ground surface or along the face of the hypothetical wall and when the wall is moved laterally in such a way that the yield condition is reached simultaneously at all depths below the surface.

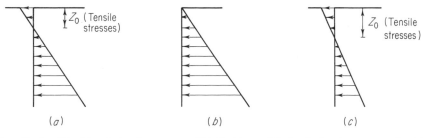

FIG. 6–4 *Limiting lateral pressure distribution with depth (active pressure). (a) General case. (b) Friction only. (c) Cohesion only.*

In Fig. 6–3 the pole of the diagram for any circle of stress at failure coincides with the intersection of the circle with the σ axis at the minor principal stress, for example, at point b'. From the pole two lines may be drawn to intersect the points at which the circle becomes tangential to the failure envelope. These lines give the orientation of the planes in the physical space along which the resultant stress obliquity has reached its maximum value, and they are termed *slip lines* or *slip planes*. The orientation of these lines may be transferred to the physical plane of Fig. 6–2 to indicate the planes on which slip is occurring when the wall yields to the left.

The slip lines are straight throughout the whole region of Fig. 6–2 because no shearing stresses were applied at the soil or wall surfaces. When, as is usually the case, shearing stresses do develop at the wall's surface as it moves, the slip lines become curved.

6–3 PASSIVE PRESSURE

It is also possible to push the wall *into* the soil until the soil yields passively; in this case the lateral principal stress becomes the major principal stress on all soil elements when surface shearing stresses are absent. When this occurs, the situation can again be shown diagrammatically in Figs. 6–5 and 6–6, which are similar to Figs. 6–2 and 6–3 for the active case. Again, the vertical stress on any element at depth z is equal to $\gamma z + q$ as shown in Fig. 6–5, but if failure is to take place as the lateral stress increases, the vertical stress now becomes the minor principal stress. Consequently in Fig. 6–6 the circles constructed for increasing lateral stress at first get smaller and then are drawn to the right at point a, which represents the stress $\gamma z + q$ as before. Considering the circle which eventually becomes tangential to the failure envelope on ab' as diameter, we see that the relationship between the principal stresses at failure expressed by Eq. (6–2) remains unchanged, but that σ_{1f}, in this passive failure condition, is now the lateral principal stress and that σ_{3f} is the vertical minor principal stress $\gamma z + q$.

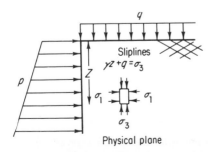

FIG. 6–5 *Special case of limiting equilibrium (passive earth pressure).*

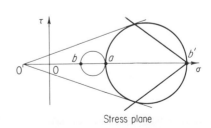

FIG. 6–6 *Stress plane of Fig. 6–5.*

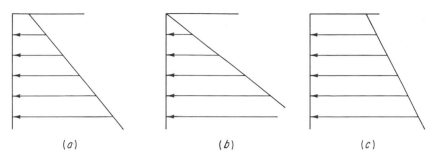

Fɪɢ. 6-7 *Limiting lateral pressure distribution with depth (passive pressure).*
(a) General case. (b) Friction only. (c) Cohesion only.

We recognize further that Eq. (6-3) will still be obtained by putting $c = 0$ in Eq. (6-2) but that, since the lateral stress (now σ_{1f}) is the unknown quantity, Eq. (6-3) is more convenient in its reciprocal form

$$\frac{\sigma_{1f}}{\sigma_{3f}} = \frac{1 + \sin \phi}{1 - \sin \phi} = K_p \qquad (6-7)$$

where K_p is referred to as the *passive earth pressure coefficient.*

When the soil possesses only cohesion, Eq. (6-4) again holds at failure, provided it is recalled that σ_{3f} is now the minor principal stress, equal to $\gamma z + q$. Consequently, Eqs. (6-2) to (6-4) represent the maximum lateral stress which can act on a wall being pushed into the soil in the manner of Fig. 6-5 under the conditions of no shearing stress at the wall-soil interface or soil surface. Under these circumstances there are no zones of tension developed in the failing soil mass, and the pressure distributions in the various materials can be shown in Fig. 6-7a to c. Under the conditions shown in Figs. 6-5 and 6-7 the maximum lateral stresses are again linear functions of depth.

From the geometry of Fig. 6-6 it will be seen that for all stress circles representing soil elements at yield, the positions of the poles will be similar, and the poles will be located at points equivalent to b' in Fig. 6-6. The slip lines in the passive case can be located by drawing rays through the point b' and the two points of tangency of the circle and the failure envelope as shown in Fig. 6-6. The two sets of parallel slip lines are also shown in Fig. 6-5, and they are again parallel only because no frictional stresses exist between the wall and the soil or at ground surface.

6-4 FAILURES BELOW FOOTINGS AND FOUNDATIONS

When the apex angle of the soil mass shown in Fig. 6-1 is greater than 90°, the distribution of stresses at failure is not as simple as is shown in the two above cases. Instead, it is found that the failure conditions inside the mass must be represented by three separate regions of yielding. An example of this situation is shown in Fig. 6–8 for the special case of interest to soil engineering in which $\alpha = 180°$. In Fig. 6–8 it will be seen that a yielding region similar to the active earth pressure situation of Fig. 6–2 lies to the left of point O, whereas another region of slip lines similar to the passive state of Fig. 6–5 exists to the right. The two zones are connected by a wedge- or fan-shaped region of slip lines. Here, one family of lines converges to point O and the other family forms arcs connecting slip lines in the two regions to each side of the fan. The situation shown in Fig. 6–8 then can be considered to represent the case of a horizontal soil surface loaded along OB with the surcharge stress q and subjected on the surface OA with a stress p sufficient to cause yielding in the soil below. The pressure p can be smaller than the stress q, in which case the problem is analogous to the active earth pressure situation, or, more commonly, the stress p is larger, causing yielding similar to that involved in a passive wall failure.

The practical problem of interest in Fig. 6–8 concerns the maximum load that can be applied to a long strip footing on ground surface (i.e., a two-dimensional yield condition is considered) before the yield strength of the soil is exceeded and the soil fails below the footing. Thus, a frequent soil engineering problem is to find the bearing capacity of a footing on top of a soil of given properties. Figure 6–8 represents a solution to this problem in the form of families of slip lines, which enable the relationship of p to q to be determined in terms of the soil properties. The solution will not be described here, but it is determined from the known conditions of yield in region BOC, which is similar to the simple passive yield state of Fig. 6–5, and in region AOD, in which yielding is occurring actively as in Fig. 6–2. The stresses in the fan-shaped zone DOC of Fig. 6–8 are intermediate or transitional yield stresses between the other two stress states. Frequently, a solution of this type of problem can be

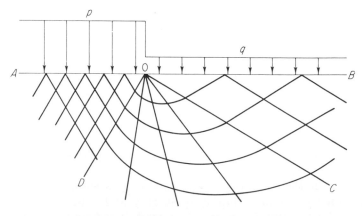

FIG. 6–8 *Special case of limiting equilibrium.*

obtained more easily when the soil is assumed to have no weight; this assumption affects the solution only when the soil possesses some angle of internal friction. When the unit weight of the soil is neglected in the solution, the slip lines as shown in Fig. 6–8 appear as straight lines and arcs for the condition of no shearing stresses at the surface. However, the slip lines become curved when the weight of the material is taken into account.

Since, as pointed out at the beginning of this chapter, the general plasticity equations are nonlinear, it is necessary to obtain a solution for every case of interest if one wishes an exact answer. Thus, the configuration of Fig. 6–8 would give rise to one value of p for a given q for a material with a certain cohesion, angle of internal friction, and unit weight and another value of p for a soil in which any of these properties were different. The correct result, precisely speaking, cannot be obtained by superposition, as with systems described by linear equations (theory of linear elasticity, for example), where the stress distribution caused by some combination of boundary stresses can be obtained by adding together the stress components arising from the boundary conditions taken separately. This is the difficulty of nonlinear problems, and because of the complexity of the mathematics, exact solutions to specific situations have not frequently been obtained.

Instead, following the techniques of linear elastic theory, conventional soil engineering practice consists of obtaining an approximate solution to a given problem by combining exact solutions for the following three separate cases: (1) the material possesses weight and an angle of internal friction but no cohesion, (2) the material has an angle of friction but no weight or cohesion, and is subjected to a surcharge q as shown in Fig. 6–8, and (3) the material has an angle of friction and cohesion but no weight. The values of the load p for these three circumstances can all be represented in the following forms:

$$p_{\phi,\gamma} = N_\gamma \gamma b \tag{6-8}$$
$$p_{\phi,q} = N_q q \tag{6-9}$$
$$p_{\phi,c} = N_c c \tag{6-10}$$

where the various subscripts on the pressure at yield p indicate the conditions under which each applies and b is the half-width of the footing. In these equations, the coefficients N_γ, N_c, and N_q all represent dimensionless earth-pressure coefficients similar to K_a and K_p but are now termed *bearing-capacity coefficients*. Each coefficient is a function of the angle of internal friction of the soil only and can be computed exactly, although with difficulty, for each of cases (1) to (3). Now we make the assumption that the yield pressure p for the general problem combining all of these cases can be calculated by adding together the values obtained for the three component problems. Thus, for a material possessing internal friction, cohesion, and weight and subjected to a surcharge q, we assume that the bearing capacity of a footing in two dimensions can be represented by the equation

$$p_{\phi,c,\gamma,q} = p = N_\gamma \gamma b + N_q q + N_c c \qquad (6\text{--}11)$$

in which the same values of the coefficients N are used as in the separate equations (6–8) to (6–10).

In this form Eq. (6–11) was first derived by Terzaghi,[1] who also obtained approximate values for each of the coefficients N as functions of ϕ. Equation (6–11) therefore represents a very approximate solution to the problem of the plastic equilibrium of a strip footing. For a few cases, the values of p obtained from Eq. (6–11) have been compared with the results of exact calculations and have been shown to be conservative [that is, Eq. (6–11) predicts a value of p at failure which is too low] by about 20 percent. When we consider the inhomogeneity and anisotropy of natural soils and the fact that the concept of an angle of internal friction and some cohesion does not describe their real behavior, the wide use in practice of Eq. (6–11) with its simplicity and convenience is easy to understand.

The variation of the three factors N with ϕ is shown in Fig. 6–9. From this figure it can be seen that the bearing capacity of a strip footing on a cohesive soil ($\phi = 0$) can be represented by Eq. (6–11) without the first

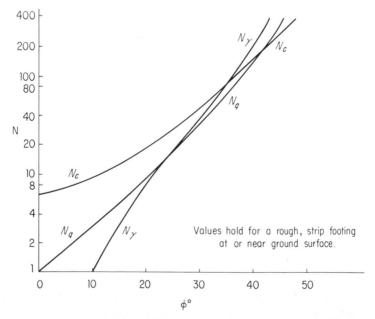

Fig. 6–9   *Bearing-capacity coefficients as functions of angle of friction.* (After Meyerhof)

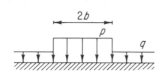

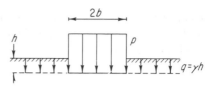

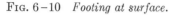

Fɪɢ. 6–10 *Footing at surface.*   Fɪɢ. 6–11 *Footing at shallow depth.*

term and in which N_c has the value of 5.1 and N_q the value of unity. When the soil possesses friction only ($c = 0$), the third term may be left out of Eq. (6–11). If no surcharge is present, the second term is omitted. Thus, the problem of estimating the bearing capacity of a footing resting on the surface of the ground which is also loaded with a surcharge as in Fig. 6–10 reduces to the determination of average soil properties over the depth influenced by the footing and their insertion in Eq. (6–11). When a water table exists in the soil affected by a footing, the buoyant unit weight is used in Eq. (6–11).

In the event that the footing, as is usual, is constructed some depth below the surface of the ground, the effect of depth can be taken into account through an equivalent surcharge q. Instead of being attributed to a surface loading, q can be used to represent the weight of soil of height h above the base plane of the footing, as shown in Fig. 6–11, so that $q = \gamma h$. In this case the bearing capacity p_h of the footing at depth h below the surface is represented by

$$p_h = N_\gamma \gamma b + N_q \gamma h + N_c c \qquad (6\text{--}12)$$

To compare the effect of placing a footing below the ground surface with putting it on the surface, we can obtain the ratio of the two bearing capacities p_h and p. Thus, for example, in a cohesionless soil we divide Eq. (6–12) by Eq. (6–11), where both c and q are equal to zero in Eq. (6–11); then

$$\frac{p_h}{p} = 1 + \frac{N_q}{N_\gamma} \frac{h}{b} \qquad (6\text{--}13)$$

For a practical range of values of angles of internal friction, Fig. 6–9 shows N_q to be approximately equal to N_γ, so that Eq. (6–13) becomes

$$\frac{p_h}{p} = 1 + \frac{h}{b} \qquad (6\text{--}14)$$

Equation (6–14) shows the rate of increase of bearing capacity with depth of imbedment of a footing in cohesionless soil. A similar expression may be written for an entirely cohesive soil by considering the special forms of Eqs. (6–11) and (6–12) required.

All of the expressions given so far for the bearing capacity of footings have referred to the two-dimensional case where the footing is long compared with its width $2b$; however, many practical cases involve footings which are circular or square, so that at failure in these circumstances the soil must yield three-dimensionally. It has been found, partly empirically but with some analytical basis, that bearing capacity for a square or circular footing can be estimated by an equation similar to Eq. (6–11) but in which the value of N_γ to be used is equal to 0.6 times the N_γ given by Fig. 6–9 for a round footing and 0.8 for a square footing; the value of N_c is equal to 1.3 times the value given in Fig. 6–9. For the square or circular footing the same value of N_q is taken as is shown in Fig. 6–9.

It will be seen that the method of considering the depth of imbedment ignores the fact that failure must also take place in the soil above the base of the footing. This limits the use of Eq. (6–12) to the case of *shallow* footings, and it is usual to consider that the equation holds only down to a depth of imbedment h equivalent to $2b$, the width of the footing. Below that level a footing is considered to be "deep" and the mechanism of failure is no longer that shown in Fig. 6–8. The problem of calculating or estimating the bearing capacity of a deep strip or circular (pile or caisson or pier) foundation is again a difficult one, and it will not be discussed here. In order to estimate the bearing capacity of such structures, values of coefficients N similar to the previous ones but taking into account the effect of depth have been approximately calculated by Meyerhof, and they may be employed in an equation similar to Eq. (6–11).

In practice, the bearing capacity of piers or piles is only approximately estimated by the use of equations, and their working capacity is usually determined by full-scale field trials of test piles or piers loaded to failure or to loads beyond the values to which the piers will be subjected in future construction.

The type of deformation illustrated by the slip-line pattern of Fig. 6–8 and applicable also to the footing cases of Figs. 6–10 and 6–11 is referred to as *general shear*, since shearing takes place in a considerable region of soil below the footing. This is the process which actually occurs in saturated cohesive soils and medium dense to dense soils. However, in loose sandy and silty soils large footing settlements can occur without a general shearing of a large region of soil. In these circumstances the soil is compressed in volume immediately below the footing and shears around the footing perimeter only. This phenomenon is called *local shear*. The mechanics of the soil deformations in this case have not been adequately analyzed, and the bearing-capacity equations (6–11) and (6–12) are usually employed; however, a modified value ϕ_l of angle of internal friction is used in determining the coefficients N_c, N_q, and N_γ, where

$$\tan \phi_l = \tfrac{2}{3} \tan \phi.$$

For footings subjected to eccentric or inclined loads, methods of estimating the bearing capacity have also been established, but they will not be described here.

6-5 SLOPE-STABILITY ANALYSES

A further failure problem of interest to soil engineers involves the stability of slopes such as those composing the sides of cuts and embankments for highways, railroads, and earth dams. Failure of such slopes can take place during construction or at periods of up to many years after construction. When an embankment fails, it is frequently found, on investigation, that a failure surface has formed in the material and that a mass of soil above this surface has slid along the surface without disturbance to the underlying soil. The acting force in these failures is due either to gravity alone or gravity acting together with earthquake accelerations, and failure develops as a result of the excessive height or steepness of the slope in relation to the soil properties. In cases where water is seeping through the soil or where pore pressures develop during the construction of an earth embankment, the reduction of effective stresses within the earth mass because of the pore pressures will frequently play a part in such failures.

No exact analytical methods for studying slope failures have been developed, and consequently most possible failure analyses are based upon the field observation of a sliding surface of failure within the earth mass. Our technique for evaluating a failure thus changes from one of attempting to solve approximately or exactly the problem of the equilibrium of yielding soil elements throughout the mass to one of considering the equilibrium of a mass of material under the action of various forces or stresses above a certain surface on which shearing may take place.

The interest in carrying out such an analysis is to determine beforehand how steep or high a slope may be constructed in a given soil. Usually a slope will not be built to the maximum configuration theoretically possible but will possess some factor of safety. The factor of safety is calculated by the analysis method for several possible slope configurations.

We have seen in an earlier section that, when a retaining wall such as shown in Fig. 6–12 moves to the left away from the soil, a minimum lateral pressure will eventually exist between the wall and the soil at yield. In this condition, slip planes may be considered to exist within the soil mass provided, as discussed earlier, that no friction exists between the soil and the wall. When the wall is of limited height, as in Fig. 6–12,

Fig. 6–12   *Equilibrium of frictionless wall.*

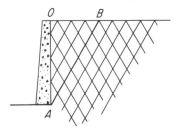

one such slip line AB may be drawn from the foot of the wall to the ground surface.

If an experiment is carried out using, for example, a cohesionless soil and conforming as closely to the required condition of no friction along the wall as possible, it will frequently be found that a wedge of soil OBA will move downward, showing that a surface of shearing or failure surface AB has developed in the soil. A wedge-shaped failure zone such as OBA will manifest itself even if friction exists between the wall and the soil and also in the case where the wall is pushed into the soil instead of being withdrawn from it. In the more usual practical circumstances, when there *is* friction between the wall and the soil, the slip surface AB in Fig. 6–12 will no longer be straight, but will be curved. However, to a satisfactory degree of approximation, the failure surface can be conveniently represented by a straight line, and a method of failure analysis consists of studying the equilibrium of the soil element OBA under the forces acting upon it. The analysis may be carried out for any type of soil.

For the general problem shown in Fig. 6–13a, where the wall surface and backfill slope are no longer vertical and horizontal, respectively, and the soil itself possesses both friction and cohesion, the potential failure surface is not given by the slip lines of figures such as Fig. 6–2 or Fig. 6–5 and the analysis proceeds as follows.

We draw any straight line AB (Fig. 6–13a) and assume that when the wall moves to the left, failure will take place as a result of movement of the wedge of soil OAB downward along the plane AB to the left and slipping with respect to the wall. In general, friction will act between the soil and the wall, and it must be taken into account in our analysis. The forces acting on the wedge OAB can be determined by studying the equilibrium of the wedge, assuming it to be just on the point of slipping. For a particular line AB the weight of the triangle of soil can be calculated

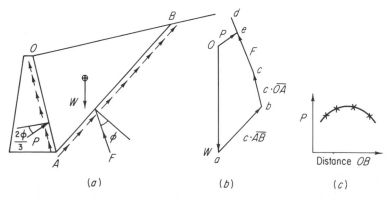

FIG. 6–13   *Force acting on retaining wall.* (a) *Forces acting on soil wedge.* (b) *Force polygon.* (c) *Maximum value of* P.

and a vertical line Oa can be drawn in a polygon of forces (Fig. 6–13b) to an appropriate scale. If the soil possesses cohesion, it may be assumed that at failure all its cohesive strength is mobilized, and this gives a force equal to cohesion c times the length of the slope AB acting parallel to AB. To the correct scale this cohesive force may be drawn as the line ab in Fig. 6–13b. Similarly, the cohesive force c times the distance OA along the wall is also known, and it may be drawn as a line bc in the force polygon of Fig. 6–13b.

Two forces are required to complete the diagram. Since the soil along the plane AB is considered to be on the point of failure, the resulting force F of unknown magnitude must act at angle ϕ, the angle of friction of the soil, to the slope AB. Consequently, from point c in the force polygon we draw a line cd parallel to the known line of action of force F. The force P acting between the wall and the soil is unknown both as to magnitude and direction, so that the determination of its magnitude requires an estimate of the value of the friction coefficient acting between the wall and the soil. It is generally taken that this cannot be greater than the friction angle of the soil itself; if it were greater, failure would take place in the soil in a plane parallel to the wall. Consequently, soil-wall friction angles equal to $\frac{2}{3}\phi$ are frequently assumed. If this is taken to be the case in our present problem, a line Oe can be drawn from point O in Fig. 6–13b at the appropriate angle to represent the force P. This line meets the force F, line cd, in the point e. The distance Oe to the scale of the figure then gives the force acting on the wall.

However, the failure plane AB was only located by guesswork, and we do not, in consequence, know that the value of P obtained is actually a maximum for this wall and soil. It is necessary to make a few more trials in the location of a failure plane AB. For three or four selected orientations of the plane AB the value of P can be computed by the method described above and plotted in a diagram such as Fig. 6–13c versus, for example, the distance OB in each case. The maximum on the curve represents the maximum value of P for a particular plane AB. This, then, is the force which would act on the wall at failure for this failure mechanism. It has been found that this method gives values in terms of an *active* earth pressure coefficient which are very close to the values that are obtained by an exact analytical solution in certain cases, but that the *passive* resistance determined by a similar method is considerably higher than that calculated analytically. This discrepancy depends on the angle of friction between the soil and the wall and occurs because of the curvature of the slip line in the exact case in contrast with the plane slip surface assumed in the approximate analysis.

It was observed as long ago as the middle of the nineteenth century that in many soils, particularly those possessing cohesion, failure in embankments and cuts took place along roughly circular arcs, and consequently a method of analysis has been developed on that basis. The method can be described as follows with reference to Fig. 6–14, which shows an embankment or cut of height H. The slope need not be a plane surface. It is desired to establish for a given height and configuration whether or not the slope will be stable for the measured values of soil cohesion, angle of internal friction, and unit weight. It is usual in these cases to determine the factor of safety of the slope.

Assuming, for the present, that the soil is homogeneous, our method of analysis is to draw a circle with center at point O in Fig. 6–14 and cutting the slope in points A and B as shown in the figure. We consider that failure may occur when the mass of soil to the right of and above the arc AB slides down the arc AB, and, as in the retaining wall problem, we analyze the equilibrium of the soil mass bounded, in this case, by the arc. From the area of the segment and its unit weight we can calculate the activating force W acting through the center of gravity of the soil mass. Since motion will take place about the center of the circle O, the weight W times the distance x of the center of gravity from the vertical through O gives an *acting moment* equal to Wx. The *resisting moment* is supplied by the shearing strength of the soil acting along the length of the arc AB. As can be seen from Fig. 6–14, this moment is equal to the shearing strength s times the arc length $\overline{AB}$ times the radius of the circle R. If the strength of the soil is known, then the factor of safety F of the

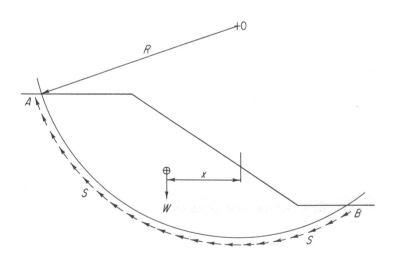

Fig. 6–14   *Stability of soil slope.*

slope for the selected circle can be calculated from

$$F = \frac{\text{resisting moment}}{\text{acting moment}} = \frac{s\overline{AB}R}{Wx} \tag{6-15}$$

If we now select a circle of differing radius or differing center or both, we can calculate a new factor of safety by the same method. The minimum factor of safety found from a succession of such trials is then approximately the factor of safety of the real slope. In practice it is found that this type of analysis yields factors of safety which are very close to those actually holding, but if the safety factor is close to unity, so that failure actually occurs, the location of the failure arc predicted by the analysis does not correspond very closely with that observed.

In the above analysis, the weight and its moment are fairly easy to calculate, but it will be recognized that a difficulty arises in summing up the resisting strength of the soil along the arc AB. If the soil consists of a natural cohesive soil and the slope is formed by excavation relatively quickly so that the pore pressures developed by the changed stress conditions do not have time to dissipate, the soil may be taken to possess cohesion only. Therefore the resistance along the arc AB is simply equal to the soil's cohesion (although different layers may possess different amounts of cohesion), measured in an unconfined compression, unconsolidated-undrained triaxial, or field vane shear test, times the length of the appropriate arc. For various soil layers, Eq. (6-15) can be rewritten in this case in the following way:

$$F = \frac{R\Sigma c\,\Delta L}{Wx} \tag{6-16}$$

in which $\Sigma c\,\Delta L$ is the summation of the cohesive shearing strength of each soil layer c times the portion of the length of the arc ΔL lying in each soil layer.

However, for calculations in cases where the pore pressures have partially or fully dissipated, the soils possess both friction and cohesion and it will be necessary to employ the full Mohr-Coulomb shearing strength equation in the analysis of the factor of safety of a given slope. In order to use this equation, the effective stresses acting along the arc must be known. In general also, the slopes involved may be irregular, and any trial failure arc will pass through materials of differing characteristics.

To take all of these factors into account, in addition to the possible presence of pore pressures along the failure arc, a variation of the method of Fig. 6-14 called the method of slices[2] is frequently employed, and this is illustrated by Fig. 6-15. For each trial circle the potentially sliding mass of soil is divided by vertical lines into a number of soil elements or slices. For the differing soil layers shown the weight of each slice may be

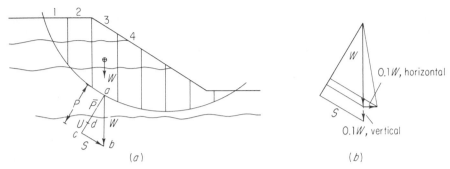

Fig. 6–15   *Method of slope-stability analysis by slices.* (a) *Typical force polygon.*
(b) *Earthquake accelerations included.*

calculated. Now, if the pore pressure conditions in the soil are known, a
force polygon can be set up for each slice such as that given for the typical slice 3 in Fig. 6–15a. As in Fig. 6–13, the weight of the slice is first
plotted vertically downward to a suitable scale starting at the bottom of
the slice; this gives line ab. At point a, a line is drawn normal to the arc
of the circle and the triangle of forces is completed by a line through
point b parallel to the tangent to the circle at point a. This enables the
components of the weight W parallel and normal to the base of the slice
to be estimated. The normal component P is given by ac in the force
diagram of Fig. 6–15a and the shearing component S by bc. If the pore
pressure at the base of the slice is known, it can be multiplied by the
length of arc traversing slice 3 to give a force U, which must act normal
to the arc.

The effective force acting on the base of the slice $\bar{P}$ is obtained by subtracting the pore pressure force U from the total force P. In Fig. 6–15a,
the pore pressure force U on the base of slice 3 is shown as cd and the
effective force $\bar{P}$ as ad. It will be appreciated that in this analysis we have
neglected the forces which must act on the sides of each slice. By neglecting them, we have tacitly assumed them to be equal and opposite in direction for each slice, but in practice these forces cannot usually be estimated.
The effects of different assumptions regarding the distribution of forces on
the sides of the slices have been analyzed by various research workers, and
it is found that their neglect does not give rise to important discrepancies
in the solution described here.

When the force polygon has been completed for each slice, the acting
moment contributed by the tangential forces S times the radial distance
R can be summed. To obtain the resisting moment the contribution of
the friction angle of the soil can be calculated for each slice by multiplying
the effective force $\bar{P}$ by the tangent of the friction angle of the soil at the
slice base and by the radius R. If cohesion is also present, it is multiplied
by the arc length and the radius R to give the cohesive component of the

resisting moment. The frictional and cohesive resisting moments can be summed for all the slices, so that the factor of safety can be calculated from the following equation:

$$F = \frac{R\Sigma c \, \Delta L + R\Sigma \bar{P} \tan \phi}{R\Sigma S} \qquad (6\text{--}17)$$

The moment arm R is the same for all forces and can be canceled from the right-hand side of the equation.

This procedure can be carried out for a number of circles until the one with the least factor of safety is once again determined. It is obvious that an incremental method like this is easily adapted for solution by a computer, and, in fact, it has been to the extent that several computer programs are available for slope-stability calculations.

Other effects may be taken into account in the incremental slice method of slope analysis. For example, the motion of an earthquake effectively increases or decreases the weight of the soil if the earthquake accelerations have vertical components and also contributes a lateral component to the weight for horizontal ground accelerations. It is seen that, for a given proportion of maximum earthquake acceleration to gravity of 0.1 or 0.2, say, the length and direction of $W(ab)$ in Fig. 6–15a can be changed appropriately as shown in Fig. 6–15b in order to give an increased shearing force for a worst possible given situation of earthquake loading. It seems likely that, under the circumstance of the short duration of earthquakes, the effective force on the base of all slices would not be increased and that the increased resultant weight would result in an increased pore pressure around the base of the circle. The circle giving the lowest factor of safety under static loading conditions is not necessarily that giving the lowest factor of safety under the dynamic conditions of an earthquake. This type of analysis of slope stability during earthquakes con-

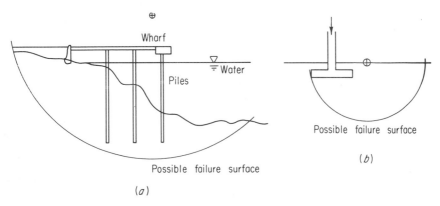

Fig. 6–16 *Other applications of slip-circle analysis. (a) Stability of waterfront structure. (b) Stability of footing in cohesive soil.*

siders that the effect of the earthquake may be taken into account by a modification of the static analysis. Much research is being devoted to the real behavior of soils and foundations under dynamic conditions in order to improve our design and analysis techniques.

The method of slip-circle analysis given here may be applied not only to slope-stability problems but also to situations such as those shown in Fig. 6–16. There are many such problems in practice to which approximate solutions may be obtained by analyses of hypothetical failure mechanisms.

PROBLEMS

6-1   Calculate the active earth pressure coefficient K_a in the case of a loose sand with an angle of friction of 30°.   *Answer*   0.33

6-2   What is the total horizontal force of the sand of Prob. 6–1 acting on a 20-ft-high frictionless wall? The sand has a unit weight of 90 pcf.   *Answer*   6,000 lb/ft

6-3   Suppose the wall of Prob. 6–2 forms part of a wharf and the water is level with the top of the wall. What is the lateral force of the sand on the wall now?   *Answer*   1,840 lb/ft

6-4   If the wall is designed for the loading of Prob. 6–3 and the water level falls, what is the effect on the stability of the wall?

6-5   If the wall of Prob. 6–2 is pushed *into* the soil, how much force is required to make the sand fail passively?   *Answer*   54,000 lb/ft

6-6   What is the depth at which the active pressure on the wall of Prob. 6–4 is zero if the soil has a unit weight of 100 psf, a cohesion of 200 pcf, and an angle of friction of 25°?   *Answer*   6.3 ft

6-7   What is the active force on the 20-ft wall (no water) for the soil of Prob. 6-6? (ignoring tensile stress down to 6.3 ft depth)   *Answer*   3,800 lb/ft

6-8   Calculate the bearing capacity (total load) of a 4-ft-wide footing on the surface of clay with a cohesion of 100 psf.   *Answer*   20,400 lb/ft run

6-9   What angle of internal friction would a cohesionless sand have to possess for the footing of Prob. 6–8 to have the same ultimate load if the soil has a unit weight of 90 pcf?   *Answer*   31°

6-10   The footing of Probs. 6–8 and 6–9 is placed 4 ft below ground surface. What is the *added* bearing pressure in (a) the clay, (b) the sand, due to the depth, if both soils have the same unit weight?   *Answer*   (a) 360 psf, (b) 9,400 psf

6-11   Give the physical reason why the effect of depth is so much more pronounced in a cohesionless soil, as shown in Prob. 6–10.

6-12   A model study is carried out on a 2-in.-wide footing on a dense sand with a unit weight of 100 pcf and an angle of internal friction of 40°. It is assumed that the sand is cohesionless, but it actually has a small cohesion of 5 psf. Calculate the bearing pressure expected for the model footing (a) when the sand is cohesionless and (b) when it has the given cohesion.

Answer (a) 920 psf, (b) 1,420 psf

6-13 Calculate the bearing pressure of a full-size 4-ft-wide footing for the soils and cases (a) and (b) of Prob. 6–12.

Answer (a) 22,000 psf, (b) 22,500 psf

6-14 What would be the danger of extrapolating from the results of the model experiment of Prob. 6–12 to the full-scale footing of Prob. 6–13 if the experimenter did not realize that the soil had a small amount of cohesion?

6-15 Use the wedge-analysis method of Fig. 6–13 to calculate the total active force acting on the retaining wall of Prob. 6–2 when it retains cohesionless soil ot unit weight 100 pcf with an angle of friction of 30°. Assume that the friction angle of the soil on the wall is equal to 30° also.

Answer 5,150 lb

6-16 Would the force on the wall of Prob. 6–15 be higher or lower if the friction between the wall and soil were reduced?

Answer About 20 percent higher

6-17 Calculate the bearing capacity of the footing and soil of Prob. 6–8 by the circular failure arc method of Fig. 6–16*b*. (Several failure circles may be tried.) *Answer* About 22,000 lb/ft

6-18 A slope of height 30 ft and slope angle 30° is built in soil with cohesion 500 psf, angle of internal friction 35°, and unit weight 100 pcf. Use the circular arc method to determine the factor of safety of the slope.

Answer About 3.05

REFERENCES

1. K. TERZAGHI, "Theoretical Soil Mechanics," John Wiley & Sons, Inc., New York, 1943.
2. D. W. TAYLOR, "Fundamentals of Soil Mechanics," John Wiley & Sons, Inc., New York, 1948.

PART 2 | SOIL ENGINEERING IN PRACTICE

With knowledge of the theory already presented in Part I, the engineer can solve all common idealized problems of soil mechanics and foundation engineering. Next he must discover means to relate his theoretical knowledge to the conditions he will encounter in practice. Frequently, he will be obliged to search for a suitable compromise between the erratic soil conditions encountered in nature and the idealized model used in the theory. This aspect of applied soil mechanics is found frustrating by some but fascinating by those who learn to combine knowledge, imagination, judgment, and experience.

Occasionally, a situation observed in practice is referred to as a "textbook case." Field trips are made and papers are written to pay homage to this unusual phenomenon. The student is left with the impression that compliance with the theories of soil mechanics is the exception rather than

161

the rule for soils as they are found in nature. This is not true, of course. Every practical problem can be solved as a textbook case, provided the conditions and the variables are analyzed and interpreted with care. At first, the engineer will be discouraged by the complexity of soil conditions on the site, but gradually he will learn to single out those properties whose variations will seriously affect his analysis. He will learn when he can safely take the average of a number of test results or when he must adhere to the least favorable results obtained. He will also discover the possibilities and the limitations of laboratory tests on small samples and of field tests on a larger scale.

The first step toward solution of any soil mechanics problem is a thorough exploration of the soil conditions in the field. This must include the geology and the topography of the site, conditions of drainage, and signs of movement or potential instability. Only after this general information has been collected and studied can locations, depths, and number of borings be selected. Samples of soil can be obtained for further analysis and testing in the laboratory.

Throughout the investigation the engineer must keep informed of the results as they materialize so he can change his original testing program to accommodate new developments. When he feels that the field results and test data form an adequate picture of the soil conditions and soil properties that exist, he must sketch an idealized soil profile with properties averaged from the test data. He then evaluates the problem by applying suitable known theoretical or experimental results, supported by his experience, to this profile before he can proceed with his report or his design. With a knowledge of the design criteria, combined with practical and financial considerations, a safe, economical design will result. This part of the book covers the steps the practicing soil engineer must take to collect his data and apply theory to a representative selection of problems.

7 | Exploration of Soil Conditions in the Field

The first step to be taken toward solution of the soil problems on a site is a careful inspection of the site and its surroundings. Unfortunately, the training of a civil engineer does little to alert or improve his powers of observation and detection. To become more skilled in drawing conclusions from the surficial features of a site, he must acquaint himself with the methods of the geologist. The geologist has been taught to look out for small details of the topography and for minor changes in the color and texture of the rock, surface soil, or vegetation. And, like a true detective, he knows how to draw inferences from seemingly insignificant facts about events which may have happened long ago. His methods of detection and deduction must be learned by the practicing soil engineer who wants to achieve competence in his profession.

The soil engineer must acquire a working knowledge of geology and must always examine available geologic data and maps at the outset of his investigation. The engineer soon discovers that geology as a science is far from exact, and initial partial knowledge of a site's geology frequently may confuse rather than inform. With the study and understanding of geologic phenomena throughout the world, a complicated jargon came into existence. The nomenclature for geologic formations and topographic features is extremely complicated, since it is based on geographical names rather than on a logical scheme. The geologic time scale is far from systematic. The terms "rock" and "stone," although precisely defined to a geologist, are used for materials which on occasion exhibit all the physical properties of a soil. The two oldest and most widely represented branches of economic geology, those concerned with mining and petroleum exploration, practically ignore the properties of the upper strata, which are referred to as "overburden." In the earth's crust these are usually the only regions in which the soil engineer is interested.

Fortunately, a branch of geology which puts emphasis on the engineering properties of the surface layers of the earth has fairly recently been recognized. This area of studies, called engineering geology, is gaining prominence with the increasing importance of understanding large-scale earth movements. It was learned by experience that nature frequently provides only a marginal factor of safety for the stability of hillsides and natural slopes. Minor railway and road cuts set off landslides involving rock and soil masses many times larger than the amount removed originally. Ancient landslides which had reached a state of equilibrium were reactivated by minor interference with the balance of forces or with the established drainage pattern. Such occurrences are now often foreseen by timely investigation by an engineering geologist. The practicing soil engineer should employ the services of an engineering geologist routinely and most particularly whenever the geology of a site appears to have a serious potential influence on his project, as is usually the case with dams, reservoirs, tunnels, and similar projects. But to understand the meaning of his expert's advice and to be able to form an engineering judgment based on the geologic conditions, he should have knowledge of the principles of

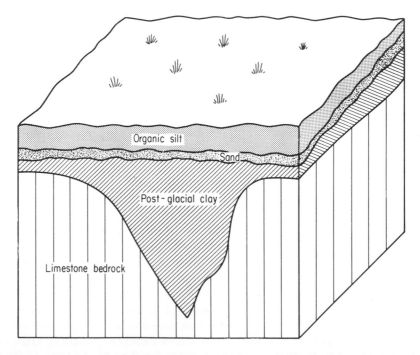

FIG. 7–1   *Buried glacial valley; 90 ft in depth, on the north shore of Lake Erie, Ontario.*

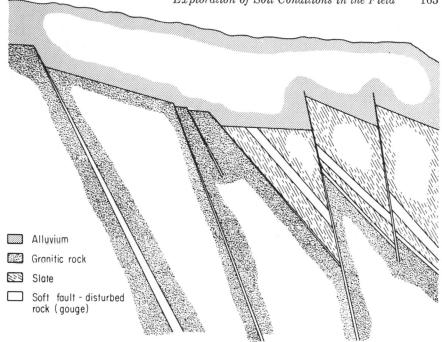

Alluvium

Granitic rock

Slate

Soft fault - disturbed
rock (gouge)

FIG. 7 – 2   *Faulting obscured by later deposits.*

historic geology and field geology as well as some experience in geologic mapping and photo interpretation. References 1 to 4 are recommended for study of these subjects. A few typical examples of frequently occurring problems in engineering geology are presented in the following paragraphs.

A foundation investigation for a structure on a terrain with relatively simple topographic features and little relief may reveal subsurface conditions which baffle the soil engineer. In glaciated country, for instance, soil deposits with a practically level surface may cover a rugged bedrock topography (Fig. 7–1). If a glaciated valley or ridge is located under the site, the geologist can provide information regarding its direction and possibly its dimensions. Additional field work to define the boundaries of such features can then be programmed with maximum efficiency. In areas of extensive folding and faulting, changes in bedrock formation in plan and in depth may be without any apparent system to the engineer and may cause doubt to be cast on the competence of this material as foundation support.

Frequently, the individual formations are quite satisfactory foundation material, but fracturing and softening along the contact planes and fault zones between formations could yield some very disturbing test data. Particularly in a small-diameter borehole, a steeply dipping fault zone could give the impression of a thick layer of soft rock flour. Figure 7–2 demonstrates such a condition as it was found below the site for a 32-story building. In this case geologic maps of the area, supplemented by careful mapping of the subterranean conditions by a geologist who observed the formations at first hand by descending into large-diameter boreholes, provided the complete picture. The disintegrated rock found in the gouge zones, although relatively soft, was well enough confined between regions of sound rock to prevent instability under the design load. In this case, where the site was located in an earthquake-susceptible area, the advice of a seismologist regarding potential seismic activity along the fault was obtained.

Whenever cuts have to be made in hillsides, conditions that promote slides in bedrock may be created. This will be the case in particular in sedimentary rock with pronounced bedding or in jointed bedrock. To determine the likelihood of instability of cut slopes, it is of importance to know the direction of potentially weak layers. This can be found by measuring the strike and the dip of the beds or of the main joint system. The strike is the direction of the trace of a bedding plane on a horizontal plane; the dip is the angle to the horizontal made by a line drawn on a bedding plane at right angles with the strike line (Fig. 7–3). This can, of course, be measured with surveying instruments, but it is usually measured with a pocket-sized instrument called a Brunton compass, which combines a compass with a spirit level. If bedrock outcrops are scarce, a number of shallow test pits will be needed. If the strikes and dips of the system as measured appear to be rather erratic, a geologist may be able to explain the conditions from the general topography and make a prediction of what conditions will be after the proposed cut is made. If the bedding dips downward out of the cut slope, or "daylights," buttress fills are often needed to support the cut. Figure 7–4 gives a typical example of what was once a steep road cut with daylighting bedding and is now reduced to the bedding plane angle by the sliding of the rock mass.

As mentioned above, similar problems can develop in bedrock whose joint system derives from stresses developed in the rock by folding and by distortions of the crust, rather than from original layering or stratification. For instance, rock at the summit of an anticline will be stressed in tension and form a system of cracks. This will develop with the years because of surface action resulting from temperature changes, wetting and drying,

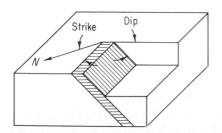

Fig. 7–3 *Dip and strike of bedding planes (strike, N20°W; dip, 35°; dip direction, S70°W).*

Fig. 7 – 4 *Slope angle flattened by bedding plane failure.*

and freezing and thawing. It will be obvious that such a joint system is far less regular than the stratification of sedimentary rock. Frequently, however, predominating directions of strike and dip of joints can be determined by an experienced geologist.

Close cooperation between engineers and geologists is mandatory for the selection and investigation of dam sites, reservoirs, underground powerhouses, and tunnels. Earth dam design and construction represent the culmination of a soil engineer's skill and the greatest challenge to his knowledge and experience. In the selection and investigation of the site, the reservoir size, ability to retain water, dam foundation and abutments, effects of a rise in water level on the surrounding land, and suitability of bedrock for powerhouse and tunnel construction are but a few of the problems that must be studied. The area involved in the investigation is extensive and frequently not readily accessible; consequently, boreholes and test pits must often be widely spaced. Geologic mapping must precede any other field work to obtain maximum benefit from a limited number of relatively costly borings.

Exploration with the aid of aerial photographs is very suitable for studies of this nature, and the techniques of photogrammetry and photogeology have greatly assisted in bringing geologists and civil engineers together. It takes considerable experience and knowledge of local soil, rock, and drainage conditions and their appearance from the air to make full use of the details revealed by aerial photographs, but study and practice of this technical aid are very rewarding and aerial photography is already indispensable for the practicing soil engineer. Rather than study a series of single aerial pictures, the engineer or geologist usually employs pairs of photographs taken adjacent to one another and involving some overlap. With suitable optical devices which range in complexity from a simple visual aid to large plotters with which contours can be mapped, the two pictures can be viewed stereoscopically to give a three-dimensional effect.

The recognition and interpretation of principal features on stereophotographs is a relatively easily acquired skill which yields valuable results. At the scales normally used for convenient and economical photoinvestigation, considerable vertical exaggeration takes place. Once this condition has been understood, it has decided advantages, as when examining the natural slopes or hillsides for stability. A slight irregularity or bulge in a slope surface, indicative of slide activity, could go unobserved on a single photograph or even in the field, but it would stand out clearly in stereovision with its exaggerated vertical scale.

Another method used to bring out special features, particularly for detail study, is the oblique photograph. In Fig. 7–5 an aerial photograph is shown; it clearly reveals the outline of a landslide which by its magnitude might have been unnoticed in the field.

In summary, the visual exploration of site conditions, both in the field and from photographs and maps, carried out with a geologist's frame of reference, will answer many engineering questions. At least it will provide a sound base on which to outline the extent and the character of exploratory work required. This will be the case in particular where existing construction provides an opportunity to check on foundation conditions in the vicinity. The engineer should never rely too heavily on local precedence, however; it is entirely possible that an uneconomical or even a technically imperfect method has been followed for years.

FIG. 7–5   *Oblique aerial photograph revealing ancient slide boundaries.*

7-2 ORGANIZATION OF FIELD WORK

After the visual inspection of the site and the study of geological data, maps, and photographs, the engineer can consider the number, type, and depth of exploratory borings required. Frequently the borings are planned too casually and without considering the need for flexibility in an exploration program. It is not uncommon for engineers in charge of soil investigations to send their field crews out to perform a fixed number of borings at predetermined locations and not to inquire about the results until all borings have been completed. This may cause considerable waste of money and effort, for the chances are high that repetitive information has been collected and some vital data are still lacking. To obtain the most useful information at minimum cost and effort, the soil engineer in charge of an investigation should follow a number of simple rules, as outlined below.

In the first place, he must know exactly what kind of information the investigation is expected to provide. This implies that he must be completely familiar with the type of construction planned on the site. If site development involving extensive earth movement is contemplated, he must have a tentative grading plan showing proposed cuts, fills, and retaining structures. If the investigation concerns a proposed building, he must be familiar with the anticipated loads on the foundations, the elevation of the lowest floor level in relation to the existing ground surface, and plans for associated construction such as driveways, planters, and retaining walls. If future extensions or additional stories are planned, he should know their locations and loads. He should know if there are special requirements regarding allowable total or differential settlements. At a bridge site, he should know the anticipated height of embankments, the number of spans, the high and low river levels, and the expected required depth of piers or foundations for scour protection.

Second, the engineer must write down minimum requirements for borehole spacing and depth and frequency of sampling while taking into consideration the type of structure and estimated soil conditions. This will permit him to make an estimate of the cost of the investigation and to schedule the field work to leave a reasonable margin for additional work that may be required. On a level, rectangular site a minimum of four borings should be considered with a maximum spacing of 100 ft. Irregularly shaped sites or erratic conditions (indicated by the first borings) will require closer spacing. Generally, a 25-ft spacing will provide adequate information for most structures even with erratic conditions. The borings should provide sufficient data to enable reliable soil profiles along two principal axes to be drawn. The minimum depth of borings should be related to the extent to which the subsoil will be affected by the proposed construction. In many cases this can be evaluated by using the concepts of stress distribution outlined in Chapter 3. A useful rule of thumb is to limit the depth of investigation to the depth at which the increase in vertical stress caused by the proposed structure will be less than 10 percent of the originally existing vertical stress.

For cuts, retaining walls, and other situations where the lateral stability is a concern, the investigation should extend below the depth of the deepest possible surface of failure. For dams, the borings should generally be deep enough below the base to give confidence in the nature, continuity, and permeability of the bedrock found. In soils the frequency of sampling is usually governed by a maximum allowable spacing of 3 to 5 ft, plus the condition that samples must be taken at each change in soil conditions. Many investigations (dams, reservoirs, landslides) demand continuous sampling, which may be obtained by special apparatus in a single borehole or, more usually, by alternating samples in closely adjacent boreholes.

Third, the engineer must select the proper drilling and sampling equipment with some attention to local preference for certain types of equipment, which exists throughout the world. Frequently these preferences make good sense: prevailing soil conditions, cost of labor and equipment, and ruggedness of terrain have aided in the development of locally suitable equipment. Other preferences cannot be defended, however, and will be abandoned eventually. In some areas, for instance,. surplus diamond-drilling equipment is frequently used for soil exploration. The wash-type borings carried out with this equipment are far from satisfactory for most soil conditions. Under given circumstances, therefore, the engineer's choice will be governed by the availability of equipment. Only very extensive investigations warrant the expense of transporting special equipment from a distant location. A summary of drilling and sampling methods is given later in this chapter.

The fourth important step is the selection of personnel to supervise the field investigation. The drilling operation is usually performed by a two-man crew: a driller and a helper. This crew will take care of the mechanics of the drilling, reaming, and cleaning out of boreholes and sampling. The more experience the crew has had in carrying out adequately supervised soil investigations, the faster the work is carried out and the better will be the samples recovered. Some drilling companies only use an experienced driller with a locally hired laborer as helper or "swamper." The saving obtained is very poor economy, since untrained help slows down the operation considerably and can be the cause of serious accidents. Since most drilling is charged by the hour for the drill plus the crew, the engineer will get the best value from an experienced team which has worked together for a long time. The field supervisor must direct the drilling and sampling, log the borings, and make decisions in the field regarding the adequacy of the information obtained. He must be able to lay out boring locations with reasonable accuracy. He must be thoroughly familiar with the drilling and sampling equipment used so he will be able to judge if progress is satisfactory and if the samples obtained are representative of the *in situ* conditions.

The problem of finding the proper personnel for this responsible type of work is one familiar to all practicing soil engineers. No schooling prepares students for this unique combination of rugged field work and delicate treatment and understanding of samples and field testing equipment, of authoritative control over the drillers coupled with the ability to observe soil and geological conditions, and of ability to write detailed and concise field reports. Often, young engineers and geologists are trained to do this work in the field after graduation, and it is certainly important that all practicing soil engineers and geologists spend at least one year in the field performing the duties of a supervisor. The most valuable field supervisors in a soil engineering company are the men who have been trained on the job, who have gained many years of experience with a variety of equipment under different subsoil conditions, and who in addition have performed all the laboratory soil tests on a range of soils.

7-3 DRILLING METHODS

The main purpose of drilling a test hole is to obtain samples of the various strata and to acquire an accurate picture of the subsoil profile at one location. Ideally, the test hole should be large enough to permit an observer to descend to log the soil profile visually. Quite often this is possible, and sometimes it is necessary. Samples obtained from inclined or distorted beds may present a misleading picture of the conditions. The steeper the inclination of the beds, the more difficult it will be to interpret the sampling results, even if an effort is made to orient the samples upon recovery. For relatively shallow explorations, test pits often reveal the cross section much more clearly than would have been possible with borings. To obtain a quick impression of the soil conditions in the upper 10 to 20 ft, a small number of bulldozer or backhoe excavations frequently offer a very economical and satisfactory solution. With this technique, care should, of course, be taken not to disturb the actual location of a proposed structure. Block samples may be taken from the excavation sides if in cohesive soils (Fig. 7–6). In noncohesive soils caving of the excavation sides limits the depth of exploration by this method. The methods of drilling a hole will be described briefly before sampling techniques are discussed.

The simplest method of advancing a borehole is the displacement boring (Fig. 7–7). This is carried out by driving a 2- to 4-in-diam sampler, which is closed off by a tight piston, into the soil and releasing the piston above the level where a sample will be taken. The sampler is withdrawn and emptied and driven again. Displacement boring is a crude method of

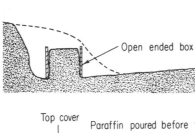

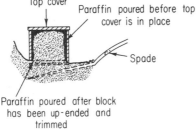

Fig. 7–6 *Block sampling of cohesive soil.*

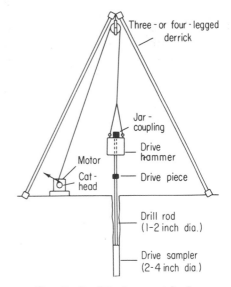

Fig. 7–7 *Displacement boring.*

continuous sampling, and it is carried on without any effort to prevent the borehole from caving or squeezing in, which may well occur in weakly cohesive or noncohesive soils. Generally, it should be considered acceptable only for preliminary reconnaissance, because the samples cannot be utilized for laboratory testing in their disturbed state. The equipment required is limited to drilling rods, a three-legged derrick, a drive hammer, and a small winch with a rope over a cathead.

Another relatively simple method of drilling is the technique known as *wash boring.* The borehole is advanced by the chopping and twisting action of a washbit, which is twisted by turning the drill rods by hand. Frequently the boring is lined with casing, which is steel pipe driven down with a large hammer. The cuttings are removed from the hole by circulating water, which is also used for jetting through the end of the washbit. The bit and rods are made to act like a plunger in the water-filled hole by moving the rods up and dropping them back by using a winch and a cathead. The equipment required, in addition to that needed for a displacement boring, comprises a pump, a water swivel, a washbit, casing, and a water tank or other source of water (Fig. 7–8).

After the hole has been advanced to a desired depth, any type of sampler may be used. However, coarse sand and gravel, if present, will collect at the bottom of the hole and form an increasingly thick barrier against sampler penetration. This may cause damage to the sampler and also cause sample disturbance. The use of water, and particularly the jetting action, affects the condition of the soil below the end of the casing. When drilling below the water table or in very impervious soils, the disturbance of the original soil caused by jetting is not very serious, particularly if the jets in the washbit are designed to spout sideways or slightly upward. Sometimes the soil profile is identified on the basis of the cutting brought to the surface by the wash water, but this procedure is very questionable, since segregation of fine and coarse particles increases with increasing depth. Only major changes in color and consistency can be observed, and sampling at close intervals is a necessity.

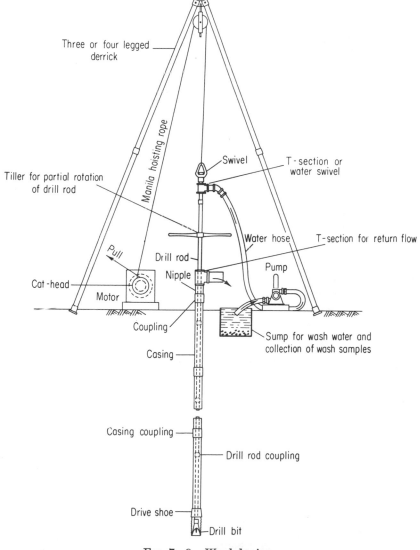

FIG. 7–8 *Wash boring.*

The sampler is lowered at the end of the drill rods, and the sample drive hammer is operated with the cathead. Often two sizes of hammer are used: the more or less standard 140-lb hammer for sampling and a heavier hammer up to 500 lb to drive the casing.

Critically speaking from the soil engineering point of view, the wash-boring method has more disadvantages than advantages. The procedure is slow, the chances of sample disturbance are great, and, on occasion, the information obtained on the changes in soil layers between samples is unreliable. The method is relatively common, however, especially in areas of extensive cohesive soil deposits. The main advantage is the limited amount of equipment needed.

Diamond drilling (Fig. 7–9) originally developed for exploration deep into bedrock, has been also adapted for drilling and sampling of soils. In rock, the borehole is advanced and a continuous core sample is obtained simultaneously by using a rotating tube or core barrel with a diamond-studded cutting edge, called a core bit. Practically all modern diamond-drill equipment is provided with hydraulic feed; i.e., a constant pressure is applied to the core barrel by fluid pressure in pistons in hydraulic cylinders as the barrel cuts its path through the rock. The bit is prevented from overheating by supplying it with water or drilling mud, which passes out into the borehole through openings in the bit. The fluid also brings the cuttings to the surface. Drilling mud is a viscous mixture of water and a sodium montmorillonite; the pressure of the clay water or mud on the

FIG. 7–9 *Diamond drill. Note core barrel and short piece of core on right side.*

sides of the hole helps to stabilize them. To protect the rock core from the erosive effect of the circulating fluid, a special metal lining is used. The whole sampling assembly is called a *double-tube core barrel*, and its use permits a very high percentage of core recovery even in highly fractured rock.

The core is twisted off the parent rock by momentarily interrupting the flow of circulating fluid, causing overheating of the bit and blocking of the rock sample inside the bit. This part of the drilling operation in particular demands skill and experience. For example, if the bit overheats too much, the diamonds may be stripped off; on the other hand, if the bit does not become hot enough, the blocking may be insufficient, so that the core drops out on withdrawal of the barrel. If the borehole sides tend to cave in when the core barrel is removed or if fluid circulation is deficient owing to the high permeability of the strata being investigated, casing must be used. As with the core barrel, the cutting edge on the casing is a diamond bit which is rotated as it is pressed into the rock. The most common sizes of diamond core barrels and casing are EX, AX, BX, and NX, which provide core sizes of $\frac{7}{8}$, $1\frac{1}{8}$, $1\frac{5}{8}$, and $2\frac{1}{8}$ in., respectively. The larger the core diameter, the greater the core strength, and better recovery and less disturbance of the cores results.

When a diamond drill is used for soil exploration, it is equipped with a cathead to drive samplers. The hole is advanced with a bit quite similar to that used in a wash boring, but instead of the constant lifting and dropping movements, the bit is rotated and fed in the same manner as a core barrel. The final result is a wash boring carried out with more expediency and with more elaborate equipment. The disadvantages of sample disturbance and lack of information between samples still prevail, and a diamond-drilling rig should not be the first choice where accurate logging and undisturbed sampling of soils are of prime importance.

A drilling method inherited from the well-drilling business uses the churn drill, which is also called a percussion or cable tool drill. Churn drills are a very familiar sight in farm country, although they are gradually being replaced by the more efficient rotary drills. Churn drilling is simply a very heavy version of wash boring; a jerk line or a crank is used to lift and drop a 1,000- to 2,000-lb string of drilling tool and rods. The amount of fluid used is considerably less than in wash boring, since no continuous circulation is required. Instead, the material ground up by the bit is churned up with a small amount of water into a slurry and then bailed out in an operation following the churning process. The slurry provides a slightly better picture of the subsoil than the segregated cuttings from wash boring. The disturbance of the soil immediately below the borehole is considerable, however, and the speed of drilling is relatively low. Churn drilling is perhaps the oldest widespread method of drilling deep holes and usually does not fit into present conditions of high labor costs.

Rotary drilling (Fig. 7–10) is in principle not different from drilling in

Fig. 7–10   *Rotary-drilling rig (Failing No. 1500).*

soil with a diamond-drill rig. The main difference lies in the size of the equipment, which is much larger on a rotary drill. In addition to the chisel-type rotary bits used on a diamond-drilling rig, roller bits are used. These bits have two to four rollers with hard-surfaced teeth which can penetrate through broken rock and boulders as well as massive rock. Very large drills of the same type are used for oil exploration. The rotary drill is used for soil-exploration work only when deep holes are required in difficult forma-tions with boulders and fractured rock or waterlogged sand. The cuttings can be brought to the surface with water, drilling mud, or compressed air. The speed of rotation is considerably less than for a diamond drill, and no water circulation is required for cooling. Drilling mud or casing will, of course, be required where there is danger of caving of the borehole sides. The size and the capacity of the power unit and the height of the mast on the drill permit the use of all kinds of sampling tools.

Auger drilling is developing as the most common method of soil exploration to depths up to about 200 ft. Most of the augering equipment in use today was designed especially for exploratory drilling or for drilling cast-in-place piles or other deep foundations. The speed of operation in most types of soil is greater than that of the other drilling methods mentioned. In many cases, the rate of advance of the hole can reach its practical optimum, which allows time for the supervisor to study the soil brought to the surface and to decide when a sample is required.

Helical augers, continuous or as a short length at the end of drilling rods, are probably the most widely used. The continuous-flight auger has the advantage of lending support to the borehole sides over the full depth of the boring. This is very useful in squeezing or caving soils, and hollow-stem augers which remain in the hole during drilling, acting as a casing, have been developed. In the latter case the end of the auger is provided with a removable plug to avoid soil squeezing up into the hollow stem during drilling. Samples are taken through the hollow stem on removal of the plug. The diameter of a continuous-flight auger is limited to 6 to 8 in. owing to the large torque required for turning a long string of augers inside a borehole. However, a short length of auger connected to strong, torque-resistant drilling rods can drill holes up to 30 in. In very stiff, cohesive soil, weathered rock, and soil containing coarse gravel and boulders, helical augers are not very effective. Very large and strong augers capable of drilling holes up to 20 ft in diameter in almost any type of soils are in use for construction of piers, caissons, and missile silos. Their use would not be economical for most exploratory work, however. A typical auger machine especially designed for soil testing and equipped with a cathead for sampling is shown in Fig. 7–11.

Particularly in western United States, bucket augers are used extensively for both soil exploration and for caisson, pile, and pier drilling. Figure 7–12 shows the details of the equipment. Bucket sizes used for drilling test holes vary from 12 to 36 in. Although progress is not as fast as with helical augers, the speed in most cases is quite adequate for soil-testing purposes. Each bucket of soil brought to the surface represents a disturbed sample from a layer whose thickness is equal to that of the height of the bucket, and this permits accurate logging. Buckets are made with a variety of different designs of teeth for scooping, scraping, and jarring soil and rock. In the deeply weathered and fractured bedrock formations prevalent in southern California, the bucket auger often can penetrate to a total of 100 ft and more. The equipment cannot be used with speed, however, when boulders larger than the diameter of the bucket are encountered. In caving soils, drilling mud can be used to stabilize the borehole sides. In this case, small amounts of mud will be brought to the surface with each bucket of soil, but this of little significance. Constant fluid circulation or jetting is not needed, which limits the soil disturbance below the borehole bottom.

A very important feature of augering large holes is the fact that the holes can be large enough for someone to be lowered down them for precise, visual logging. Naturally, the safety of personnel should always be the first consideration, and no one should be allowed to descend in an uncased hole if there is a possibility of caving or breaking loose of rocks. The observer can be lowered with the rig, and a steel-wire cage can be used for protection. In organic soils or rocks, there is the possibility of gas formation, and clean air should be pumped into the hole or the observer should wear a diver's mouthpiece with an air hose to a supply of compressed air

FIG. 7-11   *Continuous-flight auger with mast and cathead for sampling.* (Courtesy of Acker Drill Co., Scranton, Pennsylvania)

FIG. 7–12   *Rotary bucket auger.* (Courtesy of Calweld, Santa Fe Springs, California)

and a regulator. The recent development of television cameras small and rugged enough to be lowered in boreholes with a diameter of 4 in. or more may eliminate the necessity of personal down-hole inspection in some cases. Cameras are no substitute, however, for seeing the soil at close quarters and digging into the borehole side with a pick or a knife.

Generally speaking, both the helical augers and the bucket augers are very efficient tools for soil exploration. Much auger equipment with special attachments for sampling and field testing is being developed for this particular purpose.

7–4 SAMPLING METHODS

In practically all cases, sampling of soil is carried out to study soil behavior in an essentially undisturbed state. In other words, a sample is removed from the natural medium and then tested in a laboratory under conditions as nearly similar to those in its original environment as are practically attainable. The reason for doing this is that, in the majority of cases, we must accept the natural behavior of soil in the field as a basis for our design techniques, and it is therefore necessary to ascertain this *in situ* behavior as closely as possible. The act of taking a sample always disturbs the soil to an extent which must be minimized by the sampling method employed. As this fact came to be recognized, improvements in sampling techniques were effected. It is quite possible that more emphasis will be placed on carrying out tests in the natural soil in the future, since the present approach of sample extraction appears to have reached its limits of accuracy and practical applicability.

As part of a semi-infinite mass, the sample before its removal is subjected to, and is in a state compatible with, a system of stresses which is inherent in its location and the geologic history of the area. As a boring advances, this stress condition is altered in the immediate vicinity of the borehole. Immediately upon driving and withdrawal of the sampler, the stress condition of the sample alters drastically. For many soil types, this change in stress does not permanently affect the sample to an extent which has engineering significance. When the sample is placed back under similar stress conditions in the laboratory, the engineering properties of the soil can in many instances be regained, and with reasonable accuracy. It should always be realized, however, that the stress condition at the depth at which the sample was obtained is not known exactly, and consequently the relation between all laboratory experiments and actual *in situ* behavior of the soil becomes only approximate. Any further disturbance of the sample between sampling and testing must be avoided in order to preserve a reasonable confidence in the practical application of the laboratory results.

There are some instances when sample disturbance is of little importance, since only the soil type and stratification need be determined. This is the case with investigations for borrow material to be used in fills and for sand and gravel exploration. In highway work and other grading work, only a description of the material to be encountered in cuts and for use in fills is required for estimating purposes. For this *disturbed sampling* a number of devices are in use. One of the most common earlier types was the slit-tube sampler, which operated by peeling soil off the sides of a borehole. Today, auger equipment is used mainly for this type of work, and no special sampling equipment is needed. The samples obtained from the auger are large enough in volume to be used for compaction tests or other tests on remolded material.

Three principal factors influence the degree of disturbance of a soil sample: the technique of drilling, the dimensions of the sampler, and the method of forcing the sampler down. There is always some disturbance below the boreholes due to stress relief, and in a soft clay this could result in inward movement of the borehole sides and bottom. The use of drilling fluid in the hole can usually prevent this condition by maintaining the original stress conditions to some extent. Cleaning tools and jets also can cause disturbance some small distance below the borehole, but this can be circumvented by discarding the upper part of any sample retrieved. The following simple rules should be followed during drilling to avoid disturbance of the soil below the hole bottom and to assure that samples will be representative of the *in situ* conditions:

1. Proceed with an uncased, unsupported hole only in firm soils with enough cohesion to prevent caving. If in doubt, use casing or drilling mud to support the boreholes.
2. When drilling in dry or only partially saturated soils above groundwater, keep the hole dry. If absolutely necessary to prevent caving, use thick drilling mud which will not influence the moisture content of the soil.
3. When drilling below the groundwater table in anything but stiff, impervious soil, keep water or drilling mud in the borehole at a level slightly above that of the groundwater table.
4. Advance casing where possible without driving, and do not advance casing beyond the borehole depth for each sample.
5. Clean borehole thoroughly before sampling.

All samplers are cylindrical in shape. The dimensions which have influence on the degree of sample disturbance are the diameter, the wall thickness, and the length. The diameter, if greater than $1\frac{1}{2}$ in. or in excess of the size of the largest stones or concretions found in the soil, will be adequate to avoid sample disturbance. Frequently, larger samples are preferable, however, for more accurate laboratory results. The ratio of length to diameter is very important, because friction on the sides of the sampler can cause compression of the sample during driving. As penetration of a sampler into soil increases, the accumulated friction along the perimeter of the sample causes a compressive stress in the soil immediately below the sampler. This stress is equal to $\pi DL/\frac{1}{4}\pi D^2 = 4L/D$ times the average frictional stress, where L is the length of the sample and D its diameter. To minimize the compressive stress, the maximum L/D ratio should be 10 to 15 for most soils, and it may be as much as 20 for wet silts and clays which develop little friction against the side of the sampler.

The other dimension of importance is the wall thickness of the sampler. The sampler can penetrate the soil only by displacing and compressing it, so that the greater the wall thickness of the sampler, the more disturbance will take place along the internal sampler perimeter. The degree of disturbance is related approximately to the area ratio

$$A \text{ (percent)} = 100 \frac{D_{ext}^2 - D_{int}^2}{D_{int}^2}$$

where D is the sampler diameter. For well-designed samplers, the area ratio is kept below 40 percent. A number of thin-walled samplers have been developed, in particular for use in soft and easily disturbed soils. In its simplest and most frequently used form, a thin-walled sampler consists of a 2- to 3-ft-long pipe with a wall thickness of about $\frac{1}{16}$ in. and a diameter

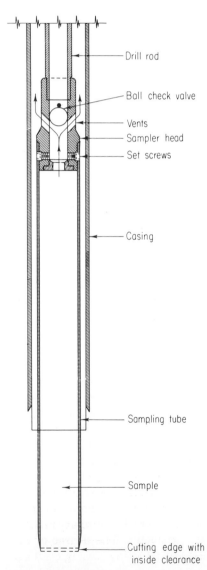

Drill rod

Ball check valve

Vents
Sampler head
Set screws

Casing

FIG. 7–13   *Thin-wall sampler.*

Sampling tube

Sample

Cutting edge with
inside clearance

Fɪɢ. 7–14　*Split-spoon sampler.*

of 2 to 3 in. At one end a cutting edge is provided; it is often made with a slightly smaller inside diameter than the tube to reduce friction against the tube walls. At the other end the tube is attached to an adapter fitted with a ball valve and vents or sometimes with a piston arrangement to provide a vacuum above the soil sample when the sampler is withdrawn. The partial vacuum helps to retain the sample in the tube. After a sample has been taken, the sampling tube can be detached from the adapter and the ends sealed off with wax for transport to the laboratory. There it may be stored or the wax plugs removed and the sample extruded for immediate testing. A section of a simple thin-walled tube with adapter is shown in Fig. 7–13.

A sampler which is very much in use in North America is the split-spoon sampler. It consists of a tube 2 in. in outside diameter and split in half longitudinally with the halves held together at one end by a threaded annulus with a cutting edge. On the other end there is an adapter with vents and a ball valve (Fig. 7–14). The split-spoon sampler is quite rugged and is used mainly in soils which are not easily disturbed or when disturbance is not important, such as dense sands and silts and firm cemented deposits. The standard method of driving this sampler is a 140-lb hammer, which is allowed to fall a distance of 30 in. This is known as the standard penetration test, and empirical data on it have been gathered since the early 1930s. For sands and granular soils in general, correlation between the standard penetration resistance (number of blows of the hammer per foot of penetration) and the relative density has been found quite reliable. When the relative density of the material is known, estimates of its angle of internal friction and the bearing capacity of a footing resting on it can be made. Typical relationships established from experience and used extensively in practice are presented in Chapter 8.

Special mention should be made of the many types of lined samplers in use. In most cases, the sampler looks very similar to the split-spoon sampler, but the size is usually larger to provide a sample with a minimum diameter of 2 in. A lining of brass or stainless steel is provided, and frequently this inner tube is divided into short sections or rings. This apparatus has the advantage that the sample can be cut to the required length in the laboratory without being taken from the rings, and laboratory equipment for routine tests such as direct shear, consolidation, and permeability can be fitted to receive the sample in the same ring. In this way the amount of handling and trimming the sample must normally be subjected to is limited, and for most soils this advantage far outweighs the disadvantage of greater overall thickness of the sampler walls. A lined sampler used sometimes in sensitive and soft clays is the Swedish foil sampler, in which the sample is separated from the sampler walls by thin metal foil which enters the sampler with the sample, thereby eliminating wall friction. Samples of 60 ft and more in length can be obtained in this manner.

Many other samplers designed for special purposes are in use: piston samplers (to further minimize disturbance), samplers for submarine work, samplers equipped with freezing elements, and samplers with cutting wires to facilitate the taking and preserving of samples in special circumstances. A detailed description of all this equipment falls outside the scope of this book; for further study reference 5 is recommended. There has been very little new development in the field of soil sampling since the early 1950s, which seems to indicate that the quality of the samples obtained with the equipment now in use is adequate and that the remaining inaccuracies in test results caused by sample disturbance are at the same level as those caused by the imperfection of the laboratory experiments.

There are several methods for advancing a sampler in soil, each of which strikes a compromise between efficient operation and minimum sample disturbance. Of these, hammering is probably the most frequently used. As mentioned before, the 140-lb hammer falling 30 in. is used in the standard penetration test. However, this blow, which is of relatively low energy, has to be repeated frequently even in quite loose soils, and this so compacts the sample that it can no longer be used for testing purposes. Use of more driving energy has an advantage in that soft or loose soils are sampled in one blow with minimum disturbance. A problem associated with fast driving is the adequacy of the vents above the sampler; particularly below water the vents must be large enough for the water to be pushed out without developing significant pressure against the top of the sample. The sampler should also be long enough to avoid its being driven beyond its free length, which would cause compression of the sample. A method which is very successful and which offers more control is pushing, in which a constant and fairly rapid rate of penetration of about 1 ft/sec is maintained.

A good method for obtaining large samples or samples in cohesive and stratified soils is hand excavation of block samples from the bottom or sides of an excavation. Disturbed samples of noncohesive soils can also be obtained in this manner, and the *in situ* density of the soil can be determined from the dimensions of the hole excavated and the weight of the sample. In the latter case the sample can be recompacted in the laboratory to obtain the same degree of density.

7-5 BORING AND SAMPLING RECORDS

In addition to maintaining firm control over drilling and sampling operations, the duty of the field supervisor is to keep accurate field records. These records must contain enough information to present the engineer in the office with a complete summary of the conditions encountered, samples obtained, field tests performed, and progress made. The field records are the most important documents of a soil investigation, and they represent information obtained at the expense of hundreds or even thousands of dollars. For his recommendations regarding foundation design, excavation, shoring, and dewatering problems the engineer relies heavily on the data presented in the field report. The report also forms the basis for his appraisal of the efficiency and competence of the drilling crew and of the field supervisor himself.

Three simple rules must be followed when preparing a field report:

1. Write down every occurrence and observation pertaining to the drilling, sampling, and testing operations and to the conditions encountered.
2. Do not rely on memory; the notes written in the field should be the final and complete field records.
3. Always place yourself in the position of the reader of the records, who was not in the field, and check if they are clear and not susceptible to misinterpretation.

When these rules are followed, even the observer with limited experience can prepare satisfactory field records. As his experience increases, he will learn to be discriminating with the information presented and to place special emphasis on certain phases of particular importance.

Practically every organization has developed its own forms of drilling and sampling records to suit its own methods of work. An early tendency toward elaborate forms with separate columns for every possible type of information that might present itself has been reversed more recently toward logs with a minimum of space allocated to specific data, a flexible depth scale, and ample space for general observations. A good field report should contain the following data:

1. Identification, date, time, place, borehole location, surface elevation, drilling equipment used, and names of driller and supervisor
2. Description of soil profile over full borehole depth, with classification of soils in accordance with a recognized classification system. Description should include details of incidental encounters with subsurface materials such as boulders, roots, and fill (drains, telephone, electric cables, gas mains)
3. Description of technique used for advancing and stabilizing borehole, if other than usual, such as the use of casing and drilling mud or methods of chopping, jetting, or blasting through obstructions
4. Listing of samples obtained, the depth range, the percentage of sample recovery if other than 100 percent, and the penetration resistance encountered
5. Rate of progress for various phases of work and comments regarding the efficiency of men and equipment
6. Report on any testing of the soil in place
7. Reliable groundwater and surface drainage observations
8. Discussions with owner, architect, or engineer visiting the site

Groundwater observations are perhaps the most frequently misinterpreted information in drilling reports. The use of water during drilling, the use of casing, the penetration of strata of vastly different permeability, the occurrence of perched water, and the limited time available for observation all contribute to the difficulty of obtaining reliable groundwater data. To begin with, it should be appreciated that, after disturbance, considerable time is required in most soils for the water table in a borehole to reach a stable level. In clays, the steady state may never come about; the permeability is so low that evaporation may remove all water appearing at the borehole perimeter, and capillary action tends to prevent seepage. In sands, the borehole may cave too badly for accurate observation. The engineer should consider how important it is to obtain a precise determination of the water table for the project under investigation before deciding if it is worth the considerable effort. In stiff clays, for instance, the water level generally does not affect the method of construction and excavation or the design of the substructure of a building. In sands and silts, the presence of a water table above the level of the proposed foundations or above proposed basement level drastically influences the design and the construction procedures. Fortunately, in those soils the water table can usually be readily distinguished as the boring is taken down, particularly if no water or mud is used while drilling above the water level.

If groundwater observations are of great importance and the soil has a relatively low permeability, the time lag involved in stabilization of the water level in the borehole may be shortened by progressively filling the borehole with water and emptying it until a falling water level changes into a rising level. For more accurate determination, a perforated casing, a wellpoint, or a porous piezometer point can be left in the borehole for future observations. Each of these devices is connected to the surface by pipe or tubing to allow frequent gaging. A seal of grout, bentonite, or other tight, impervious material is placed above the device to avoid surface-water penetration. Observations over a considerable length of time are sometimes useful to determine the variations in the water table with the seasons and during periods of high precipitation.

7-6 TESTING OF THE SOIL IN PLACE

Load Tests. Long before methods had been developed to sample the soil and to determine its engineering properties in the laboratory, efforts were made to test the soil in place. One of the simplest tests of this kind is the load test, which for a long time was considered the most practical and reliable method for determining the allowable bearing capacity of a soil.

A typical arrangement for load testing is shown in Fig. 7–15. The load is applied to a square or circular plate which is placed on the undisturbed

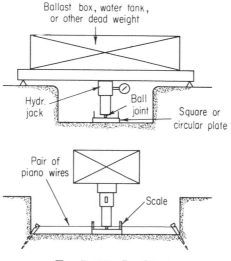

Fɪɢ. 7–15 *Load test.*

soil at the level to be tested. To eliminate the influence of the soil above testing level, this soil is removed from an area with four times the width of the test plate. The test load is usually applied by jacking against a dead weight which is supported at the surface. The load is applied in regular increments, and each load is maintained until the settlement has ceased or until failure is evident from continuing settlement at a constant or increasing rate.

The discrepancies in performance between test loads and critical footings with dimensions well in excess of those of the loading plate convinced many practitioners of load testing of the complexities of soil mechanics. The theory of stress distribution in an isotropic, semi-infinite elastic medium (see Chapter 3) can give a relationship between footing dimensions and settlement for footings with identical unit loading on such a medium. However, situations where such a relation can be used without modification rarely occur in soil engineering. In addition, the elastic properties of natural deposits may vary with depth and location, and a large, full-scale footing will stress deeper layers which may have been left practically unaffected by a test load on a small loading plate. Add to this the complications of nonlinear soil behavior, long-time consolidation process, and plastic deformation and it becomes obvious why load tests have lost popularity.

The American Society for Testing Materials (ASTM) has prepared specifications for load testing, suggesting that the plate size should not be less than half the size of the largest footing contemplated. This eliminates much of the uncertainty of extrapolation of results of test loads on small surfaces to the bearing properties of large footings, but it makes testing for substantial foundation loads prohibitive in cost. Since the factor of safety against failure of soil under a foundation is normally about 3.0 or more, the test load on a plate one-half the size of a proposed footing would be three-fourths of the design load. As a rule it is not economically feasible to perform load tests of this magnitude.

There are, however, conditions in which test loading in the field provides a good solution for the prediction of *settlements*. For this purpose the load plate size of one-half that of the footing is generally adequate, and the test pressure need not be taken greater than the design pressure of the footing, limiting the test load to one-fourth of the design load. Settlement predictions based on laboratory tests of soil samples, and soil profiles from field exploration, have a relatively low degree of accuracy, with a magnitude of error of plus or minus 50 percent. For settlement-sensitive structures, such as a continuous, multiple-span bridge truss, the accuracy of such settlement predictions may not be adequate. An example of this situation was experienced at the site of a highway bridge across the railway switchyard near Toronto, Canada. The most economical design for this bridge was thought to be a continuous concrete structure with sup-

Fig. 7–16 *Pile load test on timber pile. Loading frame, anchored to four concrete piles, provides test load reaction.* (Courtesy J. H. Baxter & Co., Los Angeles)

ports at relatively short (32-ft) intervals. Conventional settlement calculations based on borings and laboratory results indicated settlement (immediate settlement and consolidation) of $\frac{1}{2}$ in. plus or minus $\frac{1}{4}$ in. The top and bottom reinforcing required to accommodate this hypothetical differential movement was quite substantial. Test loading on a fairly large scale permitted a refinement of this settlement estimate to plus or minus $\frac{1}{10}$ in.

Another form of load testing which is still very much in use is the pile load test (Fig. 7–16). The problem of accurate determination of pile bearing capacity, particularly for piles with combined friction and end bearing, is mentioned in Chapter 8. Pile load tests are common on most large piling projects, notwithstanding their relatively high cost. The load is applied through jacking against a dead weight or by using two adjacent piles as anchors, as shown in the illustration. As with the footing load test, the ultimate bearing capacity of a pile is 3 or more times the design load, and testing piles to failure is rarely done. As a rule, tests are stopped at 1.5 to 2.0 times the design load. Although this procedure is understandable from the point of view of the cost of the test, it is regrettable in that the true factor of safety (and hence the possibility of considerable waste) is not revealed.

Dynamic Penetration Tests.   Dynamic penetration tests include various types of devices which are driven into the soil with a hammer or a drop weight. The end driven into the soil may be flat or conical or may be a split-spoon sampler. The driving resistance in blows per foot of penetration is plotted against depth, and the resulting diagram gives some indication of the changes in density and soil type. Such a graph is of very little value for the determination of the engineering properties of the soil; it merely serves, when combined with test borings, to locate distinct, relative changes in subsoil conditions such as the transition from soft clay or peat to dense sand or gravel. Although many efforts have been made to correlate the penetration resistance of a cone with relative density and ultimately with the bearing properties of a soil in the same manner as in the standard penetration test (see page 204), none has been successful. One reason for this failure is the increasing effect of soil friction on the shaft as the penetration increases, but the main problem is that the results of all dynamic or impact penetration tests, including the standard penetration test and all pile-driving tests, cannot be correlated adequately with the behavior of soil under static conditions. The applicability of the results of the standard penetration tests to practical problems is based solely on the wealth of statistical information collected and not on an understanding of the soil behavior during the test.

Static Cone Penetration Test.   A typical static cone penetration device and a penetration diagram are shown in Fig. 7–17. The casing with

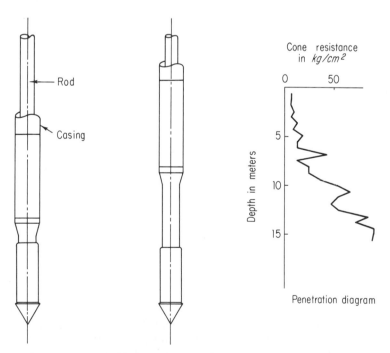

FIG. 7–17   *Static penetration cone.* (Courtesy Delft Laboratory of Soil Mechanics, Netherlands)

the cone at the end is pushed into the soil by hand, mechanically, or hydraulically. At any required depth the cone can be pushed ahead of the casing by the rod, which moves almost frictionlessly inside the casing. The force required to push the cone about four inches is measured and represents the failure load on the cone at this particular depth, independent of the casing resistance. Hence in principle the static cone penetration test is a load test on a very small scale, and the limitations regarding settlement predictions from load tests also prevail for the static cone penetration test. There are, however, two points of difference which increase the usefulness of the static cone penetration test in comparison with the load test. In the first place, the test can be performed virtually continuously over the entire depth profile which may be of interest, and in the second place, the test is carried to complete failure of the soil rather than to some arbitrary amount of strain as in a load test. Thus, provided the numerical influence of the cone dimensions on the test results can be established,[6] the results can represent the ultimate bearing capacity of the soil at closely spaced intervals. Separate tests may also be performed closely adjoining one another laterally.

In Chapter 6 it was shown that the ultimate bearing capacity of a foundation can be expressed with the bearing capacity factors N, N_c, and N_q. With increasing depth-diameter ratio of a footing, pile tip, or cone point the factor $N_q \gamma h$ increases in importance, whereas the factor $N \gamma b$ decreases to a negligible fraction of the total bearing capacity. Consequently, the ultimate bearing capacity of a small penetration cone at some depth will be practically the same as that of a narrow footing at the same depth or as the end-bearing capacity of a pile. As a result, cone penetration test results can be interpreted directly in terms of ultimate bearing capacity for end-bearing piles or deep, small footings. As can be seen from Fig. 7-17, the small cone will record changes in bearing capacity in very small depth intervals. A larger pile tip will involve a soil mass with substantial thickness in failure, somewhere on the order of 8 times the diameter of the pile tip, extending upward from the tip. Therefore the ultimate pile bearing interpreted from cone penetration tests results is based on the average cone penetration resistance over a depth range of 8 times the diameter of the pile tip. A factor of safety of 2 or 3 is applied to this value.

In summary, the static cone penetration test, if combined with borings, can be used with confidence for the determination of the end-bearing capacity of piles or deep foundations. Other uses may be found locally, but the results must be backed up with sufficient full-scale measurements for comparison. The use of the tests is, of course, limited to soils with grain sizes which are small in comparison with the size of the cone, i.e., sands and finer soils.

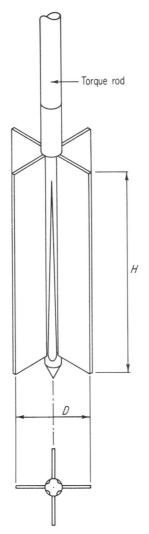

FIG. 7–18   *Field vane apparatus.*

Vane Tests. A vane consists of four thin rectangular blades welded to a thin torque rod (Fig. 7–18). The vane is pushed into the soil to be tested, and the maximum torque required to rotate the vane is measured. Usually the vane is lowered into a cased borehole to eliminate the friction on the rod, and the torque measured is that required to shear the soil along a cylindrical surface. If the end resistance is neglected, the relation between the cohesive strength c, vane diameter D, vane length H, and the torque required T is $T = \pi D^2 H c/2$. Of course, this relationship is valid only under the assumptions that the angle of friction of the soil during the test is zero and that shearing develops along the cylindrical surface.[7,8] Consequently, the vane test is employed only to determine the cohesive strength on the same basis as it is determined in quick or undrained laboratory shear tests. Elaborate torque- and strain-measuring devices have been designed for vane testing equipment, but they contribute little or nothing to a better understanding of the state of stress and the failure mechanism in the soil, and a simple torque wrench will yield the same information.

The vane may be used also to determine if a clay is sensitive to remolding. As described in Chapter 4, some clays, usually called sensitive clays, lose much of their in-place strength upon remolding, and it is important to know this if pile driving or excavation is contemplated. To determine the remolding strength, after measuring the undisturbed strength, the vane must be rotated several times to remold the soil and another torque test must be carried out immediately after remolding. The ratio between the original strength recorded and the strength measured after remolding is called the *sensitivity*. The value obtained is dependent on the methods of remolding and testing, but the vane test is a useful and simple method for comparing the sensitivity of different soils.

Pressuremeter. The pressuremeter, invented by L. Ménard, is basically a cylindrical balloon which fits snugly in a borehole and can be inflated to about twice its original diameter and thereby expand the walls of the hole (Fig. 7–19). The required pressure and volume are measured, and it can be seen that the device carries out a type of stress-strain test on

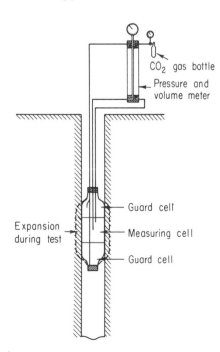

Fig. 7–19 *Ménard pressuremeter.*

the adjacent soil. To reduce the complicated end effects, the balloon is divided into three cells; the end cells expand at the same rate as the central cell, and only the central cell is used to measure the volume change. Assuming a linearly elastic behavior of the soil and a constant value for the Poisson's ratio, a value for the Young's modulus E can be determined from the test results. As the pressure is increased, the soil begins to deform plastically, the rate of strain increases, and a thickening annulus of plastic soil will extend outward from the cell. The cohesive strength of a saturated clay can be determined from the conditions when plastic deformation begins, but determination of other shear strength parameters becomes very complicated. At this time, the pressuremeter is of little more use than the vane test for determination of the shear strength of soils. In the "elastic" range of deformations it offers a rapid method for determining the Young's modulus, a method which is more applicable to practical conditions than the seismic methods in use.

7-7 GEOPHYSICAL METHODS

The term "geophysical methods" is used rather loosely by soil engineers to encompass all tests which make use of certain gross physical properties of the subsoil, such as electrical conductivity or velocity of elastic wave propagation, to obtain information regarding the subsoil profile and the mechanical properties of the soil. Most methods in use originated with prospecting for oil and minerals and are of only limited use to the civil engineer, since the information provided is too general. For instance, a seismic or sonic sounding (as described below) may reveal that an area is covered with 25 ft of relatively loose soil underlain by denser material. The type of material and its properties pertaining to settlements or bearing capacity must still be determined by other means. The techniques are useful, however, once a few widely spaced borings have established the constituents and the properties of various strata, for the purpose of making a quick and relatively accurate interpolation between the borings. In this way valuable preliminary information can be obtained in an area on the extent of possible borrow pits or the amount of a given material requiring a specific method of excavation. Of the geophysical methods in use, only three methods find some degree of acceptance for use in soil engineering practice: electrical, seismic, and radioactive methods.

Electrical Methods.   The electrical method which has found wide use for civil engineering purposes is the resistivity method. Electrical current is supplied to the soil from two *current electrodes* and the potential drop between two *potential electrodes* is measured, from which the resistivity of the soil can be determined.

A diagram of the most common four-electrode arrangement is shown in

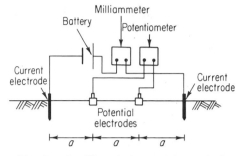

FIG. 7 – 20 *Electrical resistivity method.*

Fig. 7–20. The electrodes are moved to successive positions, and the resistivities obtained can be related to the soil profile when some information on the soil types and resistivities present is available. Empirical data about the resistivity of many soil types under different groundwater conditions have been collected. As a rule, sand and gravel have a resistivity which is hundreds to thousands of times higher than that of most silts and clays; weathered rock has a lower resistivity than massive rock; marine deposits have less resistivity than fresh water deposits, and so on. The possibility of errors of interpretation exists in particular in glacial soils, where coarse and fine soils are thoroughly intermixed and where, for instance, water-bearing gravel may be mistaken for practically impervious glacial till or boulder clay. In areas where considerable knowledge about the principal formations is available, the chances of error are smaller, and the resistivity method has been used with success in highway work and other operations where large-scale but relatively shallow excavation is contemplated.

Seismic Methods. The seismic methods of use to the soil engineer are based on the measurement of the velocities of wave propagation in the upper strata of soil and rock. In the test, vibrations are produced in the soil or rock by detonating an explosive charge or by striking a plate on the ground surface with a sledge hammer and recording the time of arrival of the first wave at some distance with a vibration-sensitive device or geophone. The spacing between shot point and geophone is increased successively, and the travel time of the wave is plotted versus the distance. With explosives, a string of geophones is usually employed, and one shot gives a considerable amount of information. When the hammer is used, on the other hand, the geophone and recorder comprise one unit, and successive hammer blows are struck at increasing distances from the unit. Each blow therefore results in one plotted point. The wave velocity in most soils varies from 1,000 to 7,000 ft/sec for the compression waves created in the test, whereas the velocity in rock formations can be from 10,000 to 25,000 ft/sec.

A typical arrangement for shallow, seismic refraction survey work is shown in Fig. 7–21. For the geophones nearest to the source, the first wave to arrive will have traveled entirely in the upper layer, and the arrival time will give the wave velocity in this medium. If the next layer is denser material, with consequently a higher wave velocity, as often is the case, the wave will be refracted in a direction parallel to the contact plane. For the geophones at greater distance, the first wave to arrive will have traveled partly through the denser medium, because the velocity in this medium is higher. The interpretation of the results becomes much more complicated when the layers are inclined or irregular, and the method fails to give information when the density of the lower stratum is less than that of the upper layer. As with the resistivity method, the seismic refraction method is useful for rapid completion of soil and rock profiles in combination with borings or other available information.

Radioactive Methods. The radioactive methods are usually described as the nuclear methods for determination of moisture content and density (and therefore unit weight) in soils. The first, used to determine the field moisture content of soils, depends on the moderation of high-energy neutrons to slow neutrons by hydrogen atoms in the free water in the soil. The second, employed in estimating the in-place density of a soil, depends on the absorption and backscattering of gamma rays by the outer-orbit electrons of the atoms present in the soil. The amount of

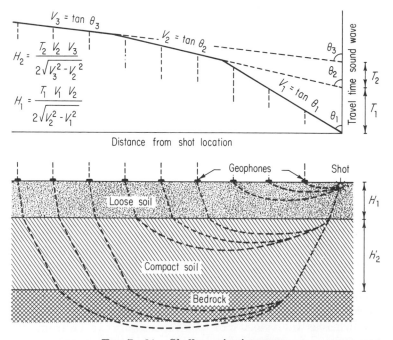

FIG. 7–21   *Shallow seismic survey.*

backscattering, measurable with a Geiger-Müller detector, is inversely proportional to the wet density of the material. Both methods involve the use of relatively delicate and expensive equipment which requires calibration in the soils to be studied, and the test results are not more accurate than can be obtained with the conventional mechanical methods described in Chapter 11. The main attraction is the speed of testing, which in some cases exceeds that of other methods. The reliability of the tests under all soil conditions is still a point of discussion.

PROBLEMS

7-1 The direction of the strike of the beds of a sedimentary rock is due north. The beds dip eastward at an angle of 30° to the horizontal. The formation will be crossed by a north-south highway, and a number of 45° road cuts will be made. Right-of-way purchases must be made, keeping in mind the potential liability of the highway department for future slides. On which side of the highway should the right-of-way acquisition be the greatest?

Answer   On the west side the beds, which are frequently potential slide planes, will dip downward out of the cut slopes; on the east side they will dip into the slopes. Sliding is more likely to occur on the west side. Land purchase should preferably be beyond the line where beds daylight.

7-2 A foundation investigation is to be made for a 100-ft-diam, 48-ft-high oil tank. The site is adjacent to a river, and ground surface is 10 ft above average river level. The depth to bedrock is known to be over 300 ft. The unit weight of the soil may be taken as 120 pcf. To what minimum depth should the soil borings be taken?

Answer   The net pressure from the oil will be about

$$0.9 \times 62.4 \times 48 = 2{,}700 \text{ psf}$$

However, the tank may be filled with water for testing purposes. The weight of the shell will be very small compared to that of its contents, so the maximum pressure at the surface of the soil will be about 3,000 psf. The investigation should be taken to a depth where the pressure increase in the soil is less than 10 percent of the existing vertical (effective) stress. This depth can be determined by using the elastic solution described in Section 3–6. A pressure bulb for circular loads is not given in this book, but for this purpose the pressure bulb for a square footing such as Fig. 8–3 is sufficiently accurate if the diameter of the tank is taken as equal to $2b$. The existing vertical effective stress at a depth h below the surface is $P_0 = [10 \times 120 + (h - 10)(120 - 62.4)]$ psf. The depth at which the pressure increase from the tank will be less than $0.1P_0$ can be found by trial and error. The required depth of boring is about 140 ft.

7-3 What would the minimum depth of exploration for the tank of Prob. 7–2 be if the groundwater level were well below 100 ft?

Answer   100 ft

7-4 A thin-walled tube sampler has an inside diameter of 4 in. and a wall thickness of $\frac{1}{16}$ in. Three-foot-long samples are taken of a saturated, medium-stiff clay. The friction between the clay and the tube is 300 psf. The unit weight of the clay is 120 pcf. (a) What force is required to push the sampler over its last fraction of an inch to obtain a 3-ft sample? (b) What is the maximum compressive stress increase on the clay at the sampler tip? (c) What is the tensile stress in the clay at the sampler tip upon withdrawal? (d) What is the area ratio of the sampler?

Answer   (a) Friction on the inside of the tube equals inside perimeter area of a 3-ft-long cylinder $\times$ 300 psf = 942 lb. Friction on the outside is slightly more, 971 lb. The resistance against the cutting can be neglected. Note that total force of almost a ton is required to push a thin sampler into a medium firm clay! (b) Compressive stress increase is $(4L/D) \times 300 = 10,800$ psf. (This is enough to cause considerable disturbance.) (c) Tensile stress = 942 lb minus the weight of the sample: sample area = 10,400 psf. This is more than enough to break the sample. (d) $A = [(4\frac{1}{8})^2 - 4^2]/4^2 = 6.3$ percent.

7-5 A foundation investigation for a 150-ft-diam oil tank in Canada consisted of four borings located at the perimeter of the proposed tank on north-south and east-west axes. The borings revealed a very consistent soil profile, with 30 to 35 ft of moderately stiff clay overlying granite bedrock. Settlement calculations indicated 6 in. of settlement under the tank perimeter and 8 in. under the center, under full load. The tank was filled, and some months later the top wall plates buckled in the northeast and southwest quadrants and the tank collapsed. The center of the tank had settled about 18 in. What could have been the cause?

Answer   The tank straddled a shallow and relatively narrow glacial valley. Geologic information revealed that all glacier movements in this region had trended roughly N40°E. The depth of clay at the valley axis was 70 ft. A better planned investigation would have included a boring under the center of the tank. Consultation with a geologist would have resulted in locating the borings on lines perpendicular and parallel to the path of glacial erosion. The consequences would have been less catastrophic if settlement observations had been made at a number of points of the tank perimeter during the initial filling period. The settlement differences would have become obvious before being large enough to cause damage.

REFERENCES

1. F. H. Lahee, "Field Geology," 6th ed., McGraw-Hill Book Company, New York, 1961.

2. "Application of Geology to Engineering Practices," Berkey Volume, Geotechnical Society of America, New York, 1962. Many contributors.
3. V. C. MILLER and C. F. MILLER, "Photogeology," McGraw-Hill Book Company, New York, 1961.
4. "Engineering Geology in Southern California," Association of Engineering Geologists, Special Publication, Los Angeles, 1966.
5. M. J. HVORSLEV, "Subsurface Exploration and Sampling of Soils for Civil Engineering Purposes," Waterways Experimental Station, Vicksburg, Miss., 1948.
6. E. E. DE BEER, The Scale Effect in the Transposition of the Results of Deep Sounding Tests on the Ultimate Bearing Capacity of Piles and Caisson Foundations, *Geotechnique*, vol. XIII, March, 1963.
7. "Symposium on Vane Shear Testing of Soils," ASTM Special Publication 193, Philadelphia, 1956.
8. H. J. GIBBS, J. W. HILF, W. G. HOLTZ, and F. C. WALKER, Shear Strength of Cohesive Soils, *Res. Conf. Shear Strength Cohesive Soils, ASCE*, 1960.

8 | Foundations

8-1 DEFINITION

The foundation of a structure serves to transfer to the earth the weight of the structure and its contents plus all forces, vertical and lateral, which may act on the structure. Some writers have suggested that the word "foundation" should include both the structure which transfers the load to the soil or the rock and the earth's crust itself, to emphasize the interdependence of the two. This is contrary to common usage of the word, however, and it entails the use of different and unfamiliar terms like "transition member" for what is generally known as "the foundation" to all practicing engineers, architects, and builders.

The foundation must perform its function in such a manner that it provides a margin of safety against collapse of the structure and that its settlements or displacements are below limits determined by conditions of practical use of the structure and by structural, aesthetic, and psychological considerations. Each of these requirements deserves the same attention; a structure which has settled excessively may be of as little use to its owner as one which has collapsed entirely. As a rule, therefore, the design of a foundation is developed in three steps. The first is to select the most economical foundation which will provide a reasonable factor of safety against failure of the soil under the anticipated loading. The second is to calculate the settlement which will develop under the design selected. If this settlement is in excess of the allowable settlement in accordance with the conditions mentioned above, the foundation design must be modified and checked again for stability. The third step is to check the feasibility of the foundation design and its construction against any special conditions which may affect the foundations, such as groundwater conditions, topography, and frost danger. The design need not necessarily be developed in the order just described; frequently, special conditions have such a dominating influence on the choice of foundations that they are taken into consideration from the beginning.

The most common foundation types are shown in Fig. 8–1. The first type indicated is the spread footing, which is simply an enlargement of the load-bearing column or wall which spreads the load over a greater area. The cost of a spread footing increases with increasing depth and size. When the footing dimensions become very large owing to the great load to be supported and the low bearing capacity of the soil, it is often practical to combine the support of several columns in one large pad or to support the entire weight of the structure with one continuous mat. When the depth required for adequate support becomes too great for open excavation, piers or belled caissons can be used. The pier shown is constructed by excavation inside a sheet-pile cofferdam, which becomes part of the structure. Belled caissons are usually drilled with a special bucket auger which can make a "bell" at the required depth (Fig. 7–12).

Instead of drilling or excavating, it is also possible to reach the soil layer which will provide adequate support by driving piles. Piles can be considered strictly end-bearing when they are driven through relatively soft materials into a dense or hard soil stratum or into bedrock. Frequently, however, the pile will obtain substantial support in friction from the soil layers penetrated, and occasionally the frictional resistance provides the main support. This is usually the case with cast-in-place, drilled piles, which are poured in a drilled hole. The purpose of friction-type piles is, of

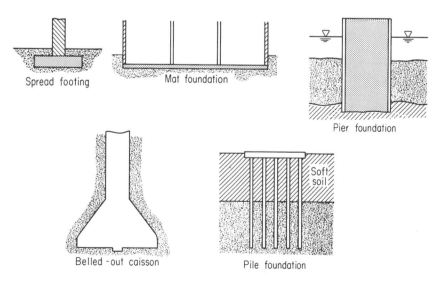

Spread footing　　　Mat foundation

Pier foundation

Belled-out caisson　　　Pile foundation

Fig. 8–1 *Foundation types.*

course, to distribute the load over a thick layer of soil which is incapable of carrying this load in direct loading through a footing or an end-bearing pile.

8-2 BEARING CAPACITY OF FOUNDATIONS

Soil Properties.   The ultimate bearing capacity of a foundation, or the load at a failure as defined in Chapter 6, is normally related to the shear strength parameters ϕ and c of the soil. These parameters are determined from laboratory tests on what are considered representative samples obtained in borings. The boring logs will give an indication of the consistency of the subsoil profile over the area of the site. Interpolation between borings involves a certain amount of speculation, and differences in soil conditions at different boreholes which do not seem to fit the generalized subsoil profile may not be ignored without a careful appraisal of their effect on the stability analysis of the foundations. If soil types, and hence soil shearing properties, vary erratically both with depth and with location on the site, the worst possible combination of high foundation pressure and poor soil strength must be investigated. One cannot take the "average shear strength" in a certain range if there is a possibility of locating a spread footing directly over the least stable soil in this depth range. Such erratic conditions often occur in flash-flood deposits, such as are commonly found in the southwestern United States, and other alluvial soils.

The situation is different when friction piles are to be used, since they extend through a large number of different soil types with variable shear properties. If the soil in a particularly weak layer is somewhat over-stressed, no pile failure need occur as a consequence, because the other layers will take over a larger portion of the shearing stresses around the pile perimeter. Consequently, notwithstanding frequent shear strength variations with depth, an average rather than a minimum shear value may be used for friction piles. Variations in this shear value over the area of the site must still be given consideration, of course.

The theory of bearing capacity of shallow footings and the more empirical determination of the bearing capacity of deep footings have been presented in Chapter 6. The parameters N_γ, N_c, N_{cq}, and $N_{\gamma q}$ can be found on the charts given in Fig. 6–9 once the appropriate angle of friction for the soil has been determined in the laboratory. Because foundations as a rule are loaded quite slowly during construction, in all but very slowly consolidating soils a drained condition may be assumed under individual footings with the pore pressure equivalent to the existing hydrostatic pressure. If a rapid increase of stress below the foundation is a distinct possibility, such as from seismic or wind loading or from filling of grain silos or water or oil tanks, the appropriate bearing capacity must be investigated in the laboratory on the basis of undrained shear tests.

Before the laboratory test results can be used with confidence for recommending allowable bearing pressures for foundations, the continuity of the parameters during the lifetime of the structure must be investigated. As a rule, vertical pressure in the soil resulting from foundation loads will tend to consolidate the soil and hence increase its shearing resistance with time. The phenomenon demonstrated in Fig. 5–15, indicating loss of shear strength during long-time shear deformations, is of little importance for horizontal foundations with horizontal bases on level land. It could have serious influence, however, in cases where the shear stresses near ground surface are high in relation to the vertical stress, as under eccentric or inclined loads or where footings are founded on a slope. Such conditions should be investigated on the basis of long-time strength predictions in a manner like that used in slope-stability analyses.

A similar problem exists with partially saturated cohesive soils. Negative pore pressure resulting from capillary suction will cause very high shear strengths, which are sometimes increased further by cementation caused by salt residues from the evaporated water. The high shear-strength values can be used safely only if there is certainty that the dry condition of the soil will be maintained throughout the lifetime of the structure. Particularly in dry, porous soils which are prevalent in arid regions, wetting of the soil by irrigation or even by a plumbing leak can have catastrophic results. If this possibilty cannot be ruled out, the soil strength should be examined by a shearing test performed on it in a saturated condition, and the effect of saturation on the consolidation characteristics should also be investigated.

Empirical Methods.   In addition to the theoretical approach to bearing-capacity problems, many empirical techniques are employed in order to determine the allowable bearing capacity on the basis of the classification of the soil, previous local experience, or field tests. For instance, most local building codes contain a section in which maximum allowable bearing capacities for different local soil types are laid down, notwithstanding the obvious risk involved in making such recommendations on the basis of visual (and often unskilled) observations of the soil. These tables are supported by local experience, and they are generally quite conservative. The recommended values are frequently employed satisfactorily with standard types of structures. A widely used empirical method is the standard penetration test, which was described in Chapter 7. The standard penetration resistance in blows per foot has been correlated with allowable bearing capacity for footings and piles in cohesionless soils. A typical set of values is shown in the table on page 204. N is the standard penetration resistance in blows per foot, and q_{allow} is the commonly used allowable soil pressure for footings up to 5 ft wide in dry, sandy soil. These values are based on allowable maximum settlements of the order of $\frac{3}{4}$ to 1 in.

N	q_{allow}, psf
10	1,500– 2,000
20	3,500– 4,000
30	6,000– 7,000
40	8,000–10,000
50	10,000–12,000

Similar relationships have been plotted in Europe for static cone penetration results. The latter are theoretically more acceptable, particularly in the fine granular, uniform alluvial deposits of the Low Countries.

Foundation Dimensions. The soil properties of unit weight, cohesion, and angle of friction may be considered as constants affecting the bearing capacity of a foundation in given circumstances. The lateral dimensions and the depth of a foundation are the variables with which the engineer must try to match the structural requirements with the properties of the soil. Equation (6–12) and Fig. 6–11 provide the means of selecting the width and depth of a foundation once the total load and the soil properties are known. The equation and graphs, and similar graphs for deep foundations, reveal a number of interesting points which provide guidelines for foundation design, namely:

1. In ideally cohesive soils (saturated clays) the bearing capacity as a pressure is independent of the width of the foundation.
2. In ideally cohesive soils the bearing capacity increases with depth from $5.7c$ to $9.7c$ plus the weight of the soil removed (γh) when h increases from nil to $10b$. For greater depths the bearing capacity increases only by the amount of soil removed (γh). In other words, it is not worthwhile to extend footings in uniform clay deeper than about five times the width, since the bearing pressure gained below this depth is less than that due to combined weight of concrete and backfill added to the footing pressure.
3. In cohesionless soil, the bearing capacity increases linearly with both width and depth of the footing.

Pile Foundations. The bearing capacity of a foundation pile cannot be established accurately with the equations and tables used for other foundations. For end-bearing piles driven through very soft soil, the equations and tables for deep foundations mentioned in Chapter 6 give quite reliable results when the pile tip lies in soils with "medium" shearing strengths, i.e., in medium clays and in granular soils with an angle of friction of 30 to 35°. For piles in circumstances when resistance is developed both by side and end bearing, and when end bearing takes place only in soils where properties may be considerably changed by the pile-driving process, empirical methods based on load tests and driving resistance are used. The static cone penetration test has been described in Chapter 7 as a relatively accurate method for determining the end-bearing capacity of a pile both in saturated clays and in noncohesive soils. In very dense granular soil the small-diameter cone records greater resistance than a larger-diameter pile, and extrapolation is not always safe. In soils with both cohesion and internal friction the cone test results cannot be interpreted reliably. If the layer in which end bearing is contemplated appears to be quite uniform, extensive laboratory testing to determine the shear propties and use of the equations for deep foundations will give reasonable results, but test loading remains the best method to arrive at a realistic value of the bearing capacity where the static cone test cannot be used.

For determining the bearing capacity of friction piles, the shear strength of the soil is measured in laboratory shear tests under normal overburden pressure. A reduction coefficient, depending on soil type, pile type, and method of pile installation, is applied to this shear strength to arrive at the perimeter shear along the pile surface. Typical values of this reduction coefficient are given in the following table.

PILE	COEFFICIENT
Drilled cast-in-place piles in nonsaturated soil	0.7–1.0
Drilled cast-in-place piles in saturated soil	0.5
Driven straight closed-end piles	0.6–0.9
Driven tapered closed-end piles	0.7–1.0

For piles driven through sand, the coefficient may become greater than 1.0 when densification due to pile driving improves the soil frictional properties. In saturated, sensitive clay, the remolding due to pile driving may cause a permanent loss of strength.

Another method for determining a pile's bearing capacity is the use of a pile-driving formula. The number of pile-driving formulas proposed could fill an entire book, and their multitude may be an indication of their limited usefulness. Most of the formulas are derived from considerations of the work done by each blow of the hammer in advancing the pile, and they attempt to calculate the resistance of the pile to penetration from the distance moved by the pile under each blow. The formulas therefore depend on a number of assumptions. The basic information used in the pile formulas is the "set," or the pile penetration per blow when penetration is nearly complete; the driving energy, and some other variables which depend on the hammer, the pile material, and the soil. Since practically every local building code prescribes the pile-driving formula which must be used locally, no formulas will be quoted here. In an alternative method,[1] the dynamic wave equations describing the motion of a rod in a resistive medium where the rod is struck at the end have been solved with a computer. In the application of this method to real piles and soil the main problem remains the selection of correct physical properties for the pile and the soil, since the soil resistance must be evaluated by watching computer predictions with pile performance during driving.

At the present time, the bearing capacity of piles under conditions where laboratory tests or static cone penetration tests may give unreliable results can best be determined from one or two full-scale pile load tests at a given site. The consistency of soil conditions over the area of the site can then be checked from the driving records of all the piles.

When piles are used in groups, the bearing capacity of the group need not be equal to that of the sum of the bearing capacities of the individual piles. This is the case with friction piles in particular. The stresses in the soil around each pile decrease with the distance from the pile, and adjoining piles will have partially overlapping stress zones. A number of "reduction coefficients" or "efficiency formulas" are in use and can be found in building codes and technical literature. The reduction applied to the sum of the individual pile-bearing capacities may vary from 0.99 to 0.50, depending on the number of piles in the group and the spacing between the piles.

Factor of Safety.   It has become customary to apply a factor of safety of 3.0 against the ultimate bearing capacity of practically all foundations, regardless of the method of determination of the bearing. When one considers the erratic nature of soil deposits and the inexact means of analyzing stresses and determining soil properties at hand, a factor of safety of this magnitude is justified. Such a factor is not particularly high when compared with safety factors used for building materials like brick, timber, and concrete, whose properties can be relatively closely controlled. For very simple cases under uniform soil conditions, or when only temporary structures are involved, a lower factor of safety of 2.0 or 2.5 is sometimes used.

In summary, the bearing capacity of a foundation on soil can generally be found to a satisfactory degree of approximation from the shearing strength properties of the soil. The future site use in its effect on the behavior of the soil should always be considered in order to see if changing conditions could affect the bearing properties. For instance, not only would a rise in groundwater level cause a decrease in shear strength of partially saturated cohesive soils but the effective unit weight of the soil would decrease by 62.4 pcf, resulting in a 50 percent or greater reduction of the bearing capacity in granular materials. Another possibility is the removal of all or part of the overburden alongside exterior footings when

excavation for future adjacent structures takes place. In built-up com-
merical areas it is always advisable to assume future excavation on adjacent property to a depth of at least one basement floor, or about 12 ft
below grade. A very common cause of footing failure results from excavation for plumbing after completion of the footings. This is often done in a
casual manner without reference to the engineer, and removal of support
of important foundations is not unusual.

Allowable bearing pressures frequently used for foundations on soil
(including the factor of safety) range from 1,500 to 5,000 psf in clays and
from 2,500 to 15,000 psf in noncohesive soils for most structures. For end-
bearing piles in dense cohesionless soil, values as high as 100 tons/ft^2 can
be used. In bedrock, theoretical failure analysis plays hardly any part in
calculations, although the crushing strength of rock under steel piles may
be investigated. The commonly used allowable bearing values for shallow
foundations on rock are quite arbitrary and range from as low as 4,000 psf
for highly weathered and fractured rock to as much as 50 tons/ft^2 on
sound granite.

8-3 STRESS DISTRIBUTION UNDER FOUNDATIONS

A theory of stress distribution in a soil is given in Chapter 3.
This analysis is important to an understanding of the general distribution
of stress in the soil below a foundation, but for a vertical stress distribution
only a simplification is generally justified. It was indicated that the distribution of vertical stresses below a foundation can be assumed to be
similar to that below a point load on the surface of a semi-infinite elastic
medium at sufficiently great depths below the foundation. The equation
governing this stress distribution, developed by Boussinesq, can be
expressed as follows:

$$\Delta\sigma_z = \frac{3P}{2\pi} \frac{z^3}{(r^2 + z^2)^{5/2}} \tag{8-1}$$

where $\Delta\sigma_z$ is the increase in vertical stress in the soil at a depth z below and
a horizontal distance r from the point of application of the point load P
(Fig. 8–2). The equation can be integrated to permit the determination of

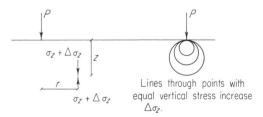

FIG. 8–2   *Symbols of the Boussinesq equation.*

stresses caused by a uniformly distributed pressure from a footing, and both graphical and tabular solutions for a variety of loading conditions have been prepared. The most common illustration for vertical stresses under foundations takes the form of a contour map of stresses, and, from the shape of the contours, gives a stress distribution called a pressure bulb (Figs. 3–8 and 8–3). For individual footings such a diagram is sufficiently accurate to enable one to determine the stress increase in a layer and to use the result for settlement calculations. If foundations are located close together, the pressure bulbs may overlap, and the combined effect of all footings at each point must be taken into account. Newmark[2] has prepared a system of influence charts with which the total stress increase at any point below a group of footings can be determined. The settlement can then be calculated for successive layers as demonstrated in Chapter 4.

Deviations of the behavior of real soils from the linear elasticity assumed in the theory does not seriously affect settlement analyses based on assumed elastic properties of the soil, provided they are measured from appropriate tests and provided unloading of the soil does not occur. The soil deformational behavior is different when stresses are decreased. A variation in the stress distribution in real soils from that given by the theory of this book occurs because soils are generally nonhomogeneous in that their modulus of elasticity increases with depth. Further deviations may occur because real soils are anisotropic. Some theoretical analyses have been carried out to give stress distributions in simple nonhomogeneous or anisotropic cases, but the results are subject to too many variables and are too complicated for practical use. They are used to give an

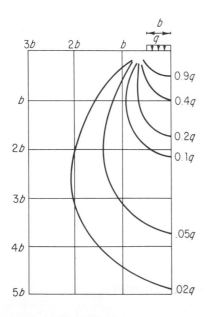

FIG. 8–3   *Lines of equal vertical stress increase below square footing on a semi-infinite, homogeneous, isotropic solid.*

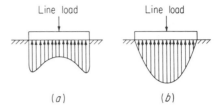

Fig. 8-4   *Vertical stress below rigid strip footing. (a) Elastic material or clay. (b) Cohesionless soil.*

indication of the differences which might be expected to rise as a result of nonhomogeneity and anisotropy.

The distribution of contact pressure immediately below the base of a foundation is affected by the relative rigidity of soil and footing. Further, yielding occurs in sands depending on the state of confinement, and there fore sand at the edge of a footing will flow plastically, offering little resistance. Sand below the center of a footing may exhibit approximately elastic behavior. A comparison of the pressure distribution below a rigid footing on an elastic material and on a sand is sketched in Fig. 8-4.

Continuous Footing.   Stress analysis becomes quite complicated when the size of the footing and the type of loading are such that the deformations of the footing itself affect the stress distribution in the soil. Individual column footings and continuous wall footings are usually so rigid compared with the supporting soil that the stress distribution depends entirely on the soil properties, and as was pointed out above, these properties can be assumed to be elastic without causing serious discrepancies in calculations. If several column loads are supported by the same footing, the lateral dimensions of the footing may become large enough that complete rigidity can no longer be obtained economically and the deformations of the footing will lead to redistribution of the stresses in the soil. The relative rigidities of footing and soil then become important.

Figure 8-5 shows the pressure distribution and beam bending moments for different degrees of flexibility of a combined footing supporting two columns. As can be seen, the bending moment in the footing decreases as the flexibility of the footing increases, and the maximum stress in the soil increases at the same time. This condition will be of even greater importance when all loads from a building are supported by the same mat or raft foundation. For such foundations a number of analyses have been made; they are based on a linearly elastic behavior of the soil material underlying the foundation. The results of such analyses may have validity in some practical cases, as, for example, the calculation of stress in railroad tracks. However, the latter are supported by a well-compacted stone bed, which may have deformation properties of a nature close to that employed in the analysis.

In most cases, on the other hand, mat foundations are selected for very

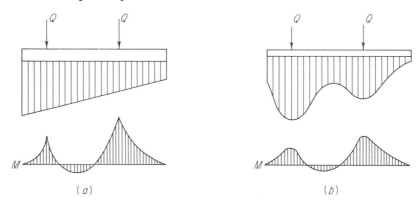

FIG. 8–5   *Pressure distribution and beam bending moments for (a) rigid and (b) flexible footing supporting two loads.*

poor soil conditions, such as soft clays which have assumed elastic properties varying in time with loading as a consequence of consolidation, and this will cause a steadily varying condition of stress below the mat with time. The consolidation will level out the high stress concentrations below the points of load application on the mat as settlement occurs at these points, but this will increase the bending moments in the mat and in the structure above it. Because it has not been practical to take such behavior into account in the design of a structure, the tendency has been to make mat foundations quite rigid and to increase the rigidity of the structure by connecting the column ends with a gridwork of ribs. Practical considerations and examples of such designs are given in reference 3.

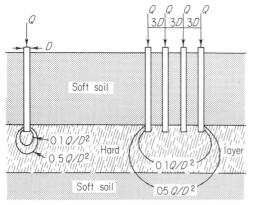

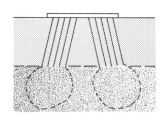

Lines of identical vertical stress increase for a single pile and a 16-pile group.

FIG. 8–6   *Reduction of pressure bulb overlap by splitting pile group into two groups of batter piles.*

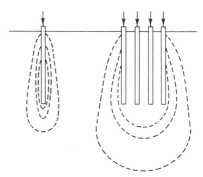

FIG. 8-7 *Pressure bulbs for single friction pile and friction pile group.*

Pile Foundations. In the case of pile foundations the stress distribution cannot be determined as simply as for spread footings, particularly not when the piles receive support both in end-bearing and in perimeter friction. Figure 8-6 shows the pressure bulb for a single end-bearing pile and for a group of end-bearing piles. The overlapping effect, which can be evaluated in the same manner as for spread footings, can be reduced by splitting large pile groups into several groups of batter piles.

With friction piles, the soil near the perimeter will receive some vertical stress over the full length of the pile. Long-time load tests on single friction piles indicate that settlement of such piles is relatively small and that complete consolidation is established in a relatively short time. This is an indication that only a small cylindrical mass of soil around the pile is actually involved in the support of the pile and that excessive pore pressure dissipates quickly in the surrounding mass. For a group of friction piles the total vertical stress in the soil below the pile tips must be investigated for settlement calculations (Fig. 8-7). With a flexible pile cap, the piles near the center of the group will tend to settle more than those at the perimeter, and if the pile cap is rigid, the perimeter piles will carry a larger than average share of the load.

8-4 SETTLEMENT OF FOUNDATIONS

As mentioned in the introduction to this section, the criteria for allowable settlements of structures are determined by conditions of practical use and by structural, aesthetic, and psychological considerations. It may seem strange at first that considerations other than those of structural safety may determine the allowable settlements. But consider, for instance, structures in use for scientific purposes such as radar domes, particle accelerators, astronomical observatories, and satellite-tracking stations. Conditions for practical use of these structures will demand fine tolerances on allowable settlements, which are probably of a different order of magnitude from those imposed by conditions of structural safety.

FIG. 8-8   *Shrine of Guadalupe, in Mexico City, has settled from 3 to 5 ft with respect to the surrounding plaza.*

A forward tilt of a high wall may be entirely acceptable from a structural or safety point of view, but psychologically such a wall is unacceptable because it creates the impression that it is likely to topple over any moment.

Uniform settlement of a structure does not affect the structural integrity in any way; theoretically, a building could be allowed to settle uniformly quite a large amount if only structural considerations were of importance. But one need only consider the effect of a uniform settlement of, say, one foot on the accessibility of a building and on the connections to the building with utilities to realize the undesirability of the situation. Naturally, it is possible to accommodate large settlements with special measures, and this is being done in many cases. In Mexico City, where builders are plagued with the problem of both overall and local subsidence owing to groundwater withdrawal and the presence of a thick layer of compressible soil, several ingenious design and construction methods have been developed. A building founded on the compressible stratum would settle at a rate well in excess of that of the general subsidence, and a building founded on piles driven into the underlying firmer soil would be left behind by the regional settlement, which at one time was as much as one foot per year (Fig. 8-8). Many important buildings are founded on what is called a *compensated pile foundation*, a system developed by Leonardo Zeevaert[4] which endeavors to maintain a rate of settlement equivalent to that of the subsidence of the surrounding land.

Apart from the problem of possible structural damage, the acceptable amount of total settlement of a factory, office, or apartment building is generally less than 3 in. For other structures the allowable settlement may be considerably more. In particular, large welded oil storage tanks have been allowed to settle considerable amounts, up to 5 ft in some instances. Here the appearance is of little importance; the structure itself is capable of sustaining relatively large differential settlements; and the pipe connections can be made flexible without difficulties. Moreover, the empty tank can be jacked up and sand injected below it quite easily if differential settlements become excessive from some point of view.

From the foregoing discussion it may be concluded that the allowable maximum uniform settlement of a structure is generally determined by considerations of practical use and by local custom. Another form of settlement which usually is not limited in the first place by structural considerations is *tilting*. Long before tilting becomes a source of possible structural damage, it will have reached the limit set by conditions of practical use or aesthetic or psychological considerations. A tilt of as little as 1 in. in 20 ft can be discovered visually by a careful observer, and a larger tilt is very objectionable.

The form of settlement remaining to be discussed, and the one which causes most distress to structures, is the *differential settlement between elements of the same structure*. In the vast majority of cases, the allowable maximum differential settlement is the determining influence on the foundation design. It does not require much differential movement between two adjoining columns of a building to cause substantial damage.

It is obvious that the maximum differential settlement between *any* two points in a building need not necessarily be the determining factor. The determining settlement is that settlement causing the maximum amount of distortion to a part of the building only. This can be expressed as the ratio between the differential settlement δ of two points and the distance between these points L, or the angular distortion δ/L. As indicated in Fig. 8–9, this does not necessarily involve the maximum total settlement or the maximum differential settlement inside the structure.

Skempton and MacDonald[5] made a survey of data concerning the settlement and damage of 98 buildings in an effort to determine a general criterion for maximum differential settlements in different buildings on the basis of the damage caused. They found that no structural damage or damage to interior finish, panel walls, or partitions occurred in the

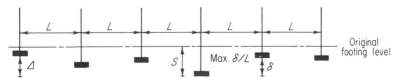

FIG. 8–9   *Maximum total settlement S, maximum differential settlement Δ, and maximum angular distortion δ/L.*

buildings investigated at angular distortions smaller than 1 : 300. In frame buildings without interior partitions or brick walls, the damage limit was about 1 : 150. This implies that for a conventional structure with brick or block partitions, a differential settlement of 1 in. in any 25-ft span can be tolerated without signs of damage. The damage limit increases substantially for very slow settlement rates. The survey was confined mainly to buildings with brick and block partitions, and little information is available about the performance of curtain walls and other types of prefabricated paneling. If adjacent differential settlements of 1 in. or more in 20 to 25 ft are anticipated, the rigidity of the frame, the partitions, and the paneling should be investigated to see if deformation-sensitive details can be avoided.

With the knowledge that differential settlements will have a dominating influence on the foundation design and perhaps even on the overall design of a structure, the soil engineer must set out to determine the settlements as accurately as possible. He must keep in mind that differences in settlement are caused not only by differences in loading but also by differences in soil conditions. The borehole and test data must be adequate to support a reasonable appraisal of the differences in compressibility of the soil on the site, both in depth and in plan.

Settlement Calculations.   Following the preliminary selection of foundation type and size on the basis of the bearing-capacity calculations, the anticipated settlements must be determined from the loading conditions and the information obtained from boring and laboratory tests, complemented sometimes by field load tests. The example of settlement

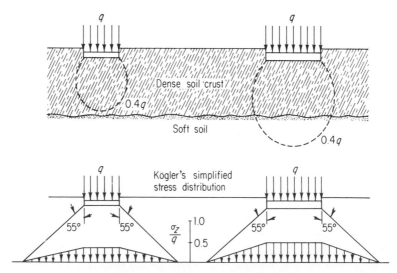

FIG. 8–10   *Strip footings on crust of dense soil overlying more compressible material.*

FIG. 8–11 *An erratic soil profile complicates settlement analysis.*

calculation given in Chapter 6 has taught us how to determine the settlement from the conditions of loading and the soil properties. The settlement values thus calculated must be studied carefully with an eye to potential differences in amounts at different locations in the site. If large differences in compressibility are found between deposits at different levels and at different positions, a tentative choice of footing level can be made with a view to bypassing the most compressible soils. If highly compressible soil is found below a "crust" of much more competent material, the effectiveness of this crust at the location where it will be thinnest and also at the point where it will be loaded most heavily must be investigated (Fig. 8–10). If the soil profile over the site area is complicated and erratic, the worst possible combination of compressible soil and high loading must be assumed (Fig. 8–11). In most cases of stratified soil the compression of each layer must be determined separately.

The magnitude and the type of loading anticipated from the proposed structure must be divided into *permanent loads*, which include both dead loading and the design or average live loading, and *short-time loads*, such as wind and seismic loading. For all but very weak soils, the latter need not be considered to contribute to the settlement of the building, and consequently, if the allowable bearing pressure is governed by settlement considerations, an increase in its value for seismic and wind loading is justified. The same cannot always be done for impact loading from other sources, as, for example, from overhead cranes, conveyors, and other machinery, because the frequent and predictable repetition of the load application may eventually cause compression of the soil in time, particularly when the soil is cohesionless. In some cases vibratory loading can cause greater settlements than static loads owing to compaction of granular soil.

The agreement between settlement calculations based on laboratory tests and actual settlements observed in the field is often not very impressive. This problem has been the subject of much research, and the differences have been blamed on such diverse factors as the inadequacy of the theoretical model, pore pressures developed under foundations, sample disturbance, rate of loading, and the high hydraulic gradient which prevails in the soil specimen in laboratory consolidation tests owing to the small height of the sample. The laboratory tests frequently produce results which are conservative; that is, the settlement in the field is smaller than that calculated. Practicing engineers often apply what is called an experience factor to reduce the calculated settlements by 30 to 50 percent, depending on local experience. This should be no excuse, however, for haphazard settlement guesses based on inadequate field or laboratory test results; the available techniques should be utilized to their fullest extent, blended with practical experience.

The time rate of settlement due to consolidation, the theoretical determination of which was given in Chapter 4, very often cannot be established accurately for field conditions. Apart from the problem of the real boundary conditions, which play an important role when the footing dimensions are of the same magnitude as the layers which are subject to consolidation, the nonconformity of the natural deposits can sometimes cause serious discrepancies. In general, the effect of nonuniform conditions is to accelerate consolidation. Thin layers of slightly more permeable soil in a clay deposit act as drains, splitting what may have been considered as a single layer into several layers. Since the time of consolidation is related to the square of the layer thickness, division of a single layer into, say, three layers by thin seams of more permeable soil will reduce the period required for 90 percent consolidation to one-ninth of a hastily assumed value! In cases where the period of consolidation is of great importance for planning purposes, test loading or settlement recording from the start of construction should supplement the data obtained from laboratory tests.

Methods for Reducing the Effects of Settlement. The fact that settlement calculations in a given circumstance indicate the likelihood of relatively large settlements does not necessarily rule out the proposed method of foundation; this is particularly true if no economical alternative is available. Many ingenious solutions have been found in cases where standard foundation techniques in an area subject to considerable settlement could not be employed. The methods used to reduce the effect of overall settlement of a structure are generally different from those used to accommodate differential settlements between parts of a structure.

One of the most frequently used methods to reduce total settlement of a structure is to preload the area to be occupied by the building. This method is especially effective on compressible soils in which consolidation occurs fairly rapidly; in slowly consolidating saturated clays of great thickness the period of preloading required to effect appreciable settlement in this manner is generally too long. Rebound may take place in some soils on removal of the surcharge before the structural weight becomes effective.

It is usually advisable to apply a preloading pressure greater than the proposed design pressure. In soils with a very porous structure and a low relative density, the compression under preloading will be rapid and very little rebound will be experienced upon removal of the load. Loosely packed and slightly cemented sands and silts fall in this category, provided the preloading pressure is high enough to break the weak bonds between particles. Bond breaking can sometimes be promoted by adding water during the preloading. Other soils which lend themselves well to preloading are organic silts and peat, in addition to sanitary land fills and other assorted rubbish, which can also be precompressed quite effectively in this manner.

Preloading is most frequently done by placing fill over the area to be stressed by the future structure, but in special cases other types of dead weight such as iron ore, stacked pulp wood, and even snow have been used. The last has the possible advantage of automatic removal in the spring, but it has the disadvantage that the effective preloading pressure cannot be determined accurately. Another preloading method used is to lower the groundwater table; this results in an increase of the effective unit weight of the drained soil of 62.4 pcf for each foot of lowering. A wellpoint system can lower the water table about 20 ft in a single stage, causing an increase in pressure of 1,250 psf on the underlying soil.

Total settlement of a structure can also be reduced by utilizing the buoyancy effect on the submerged part of the structure. By constructing a watertight basement below the water table, the uplift pressure from the groundwater can be utilized. The question arises: will the water table remain at a constant level? Unless the structure is located close to a large body of water, a constant level is not likely and the chance of a drop in the groundwater table cannot be taken. A lowering of the groundwater table, such as often takes place in areas with increasing population, frequently has the effect of general subsidence for the same reason as given in the case of preloading. The special solutions for the subsidence problem in Mexico City have been mentioned. It is also possible to reduce the net load of the structure on the ground by excavating an amount of soil for the structure whose weight is approximately equal to the weight of the building. This may be a desirable solution if the resulting basement space can be utilized.

Anticipated differential settlements between parts of a structure can be reduced by a number of measures. As a rule, because of the normal pattern of loading and the influence of adjoining footings on one another, settlement under a building will be higher at the center than around the perimeter. In a regular grid of columns a corner or side column carries only a fraction of the load on a center column. In addition, the stress distribution below a footing for a corner column is usually affected only by the stresses of two adjoining perimeter footings, whereas the stress under a center column can result from four adjoining center columns. As a consequence, the settlement of a center column may be more than four times as high as that of a corner column. By cantilevering the building out over the perimeter columns, the latter will carry as much load as or even more load than the center columns, thereby providing a more evenly distributed settle-

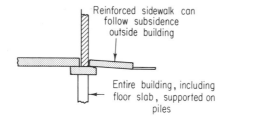

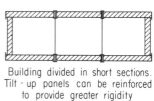

FIG. 8−12   *Practical examples of accommodating differential settlements.*

ment. Another possibility would be to combine the footings of a number of interior columns in a footing of larger area, resulting in a lower unit bearing pressure under the combined footing. This method is particularly effective if the layer causing most of the settlement is of limited thickness. For compressible layers with a thickness in excess of twice the footing width, spreading the load over a larger area has little effect on the settlement.

An entirely different approach to the accommodation of differential settlements is to provide greater flexibility between the elements of the structure while accepting the expected magnitudes of settlements. This can be done by construction in independent segments, use of appropriate construction joints, and many other techniques. Figure 8–12 shows some practical solutions for reducing the effects of differential settlements.

As a last resort, it is always possible to provide adjustable supports in the form of hydraulic or mechanical jacks, which can eliminate the differences in settlement at appropriate time intervals. For mechanical equipment in machine shops and for the supports of cranes, radar equipment, and kilns and similar equipment this is a fairly common practice, because the tolerances required are of a different order of magnitude from those which can be guaranteed by the foundation engineer. For normal building construction the problem of maintenance makes a foundation on jacks less attractive, but automatically controlled systems of jacks have been installed in special cases.

8-5   SPECIAL CONDITIONS AFFECTING THE CHOICE AND DESIGN OF FOUNDATIONS

Although in many cases the calculations of bearing capacity and settlement will furnish the most desirable and economical foundation design, there are other factors which may influence the design to a very large extent. The most important of these factors are the design of the building structure itself, the position and movement of the groundwater, the topography of the site, and the presence of adjoining structures.

Structural Design.   If a structure is designed as a multiple-span rigid frame, a continuous beam on many supports, or a similarly rigid type of construction, the allowable differential settlements become very small and structural damage rather than architectural damage may tend to occur. Obviously, a very conservative foundation design is required, and frequently such a rigid, statically indeterminate construction rules out the possibility of spread footings, because settlement differences of a magnitude of, say, $\frac{1}{8}$ in. between 15- or 20-ft centers are practically inevitable for spread footings on soil. A similar problem exists for rigid frame construction with fixed column ends at footing level. A spread footing can support the resulting bending moment at a column base, but this will cause eccentric footing pressures, and some rotation of the footing cannot be avoided. A difference in settlement of $\frac{1}{16}$ in. between the side of a 4-ft-wide footing with the highest pressure and the side with the lowest pressure will cause a substantial redistribution of stresses in a frame. For these types of structure a very large footing or a pile foundation is required.

Groundwater.   The presence of free groundwater on a site at a level near or above that of the proposed foundations presents a number of complications, in particular in relatively permeable, cohesionless soils. In sandy and silty soils, excavation below groundwater leads to considerable disturbance of the excavation bottom unless the water table is depressed locally to maintain a dry excavation for the foundation construction. The question of the excavating techniques involving dewatering will be deferred to Chapter 10. However, the dewatering procedure required may be so costly that a different foundation type, such as driven piles, may be considered preferable. Changes in groundwater level can cause consolidation of the soil due to various effects such as the increased effective weight of the drained soil, shrinking and swelling in expansive soils, and collapse of dry, porous soils upon saturation.

Expansive soils are considered to be clays or organic silts which exhibit a relatively large (5 percent or more) volume change when the soil is saturated or dried, even when external pressures are applied. Large structures are usually not founded on such soils unless the soils are dry and the load of the structure is adequate to prevent swelling when moisture is added to the soil. For house construction on swelling soils, measures are taken either to induce expansion before construction or to limit it. This is done by soaking the soil before construction, by providing a vapor barrier of polyethylene under the house, and by landscaping with short-root vegetation requiring frequent irrigation around the house. For residential construction the risk of future shrinking of the soil is sometimes considered acceptable, particularly when measures are being taken to maintain a moist condition under the houses.

Another condition affecting the selection and design of foundations, indirectly caused by groundwater conditions, is the possibility of freezing and thawing of the soil below the foundations. Dry soils experience very little volume change during a freezing and thawing cycle, but the proximity of free groundwater can create very serious problems. In what are called frost-susceptible soils, the volume increase during freezing may be many times more than that which could be expected from a volume increase of the pore water when it changes into ice. This can be explained by the fact that such soils, which range in grain size from silty sands to clayey silts, draw in water from the underlying soil or water table as they freeze, causing the formation of *ice lenses* (Fig. 8–13). The development of ice lenses, which results in heaving of the soil, is conditional upon the presence of water adjacent to the freezing soil. For many years use was made of a criterion which considered a soil to be susceptible to frost heave when it possessed 3 percent or more by weight of grains passing the No. 200 (0.074-mm) sieve. Further studies[6] have shown that the mineralogical and chemical composition of the fine fraction of the soil determines the dividing line between non-frost-susceptible and frost-susceptible soils. In particular, the presence of the mineral kaolinite and that of metal hydroxides and organic matter contributes to the phenomenon of frost heave.

In moderate climates, foundations are protected against the effects of frost simply by placing the foundations deep enough to avoid frost heave below the footing level. The maximum depth of frost penetration in the northern United States (outside Alaska) can be as much as 8 ft, and greater depths are reached farther north, where finally a condition of permanently frozen soil develops. In the latter areas it is not the frost penetration, but the thaw penetration in the summer which causes concern, as will be described later.

Foundations do not necessarily have to be taken down to the maximum known depth of frost penetration to avoid damage. In coarse, sandy soils and in rock there is no danger of excessive volume changes under freezing. In addition, the maximum depth of frost penetration will rarely be found near heated buildings, and a depth of two-thirds or three-quarters of the maximum recorded depth of frost penetration is generally quite adequate. Local experience is the best guide, of course, and many local building codes specify minimum limits for foundation depth to provide frost protection. On the other hand, the solution employing deeper foundations does not completely eliminate the possibility of frost damage to a structure. In the

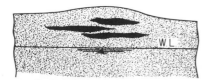

Fig. 8–13 *Formation of ice lenses in the capillary zone causes frost heave.*

case of unheated buildings or garden walls, it is possible that the entire structure will be lifted off its foundations owing to swelling and heaving of the surrounding soil. If such a structure is planned in a soil which is particularly susceptible to frost heaving, a coarse, loose backfill can be used around the sides.

The problem of freezing and its effect on foundations must be especially investigated for the construction of ice rinks, cold-storage facilities, and liquid natural-gas storage facilities, where continuous subfreezing temperatures may cause very deep frost penetration into the soil. The solution for these structures is usually found in adequate insulation between the cooling system and the subsoil, rather than in deep foundations. In the case of liquid-gas storage, the freezing of the soil is deliberately induced and the frozen ground itself becomes part of the retaining structure for the liquid. Here considerations of heaving and the strength of the frozen soil are important.

In the far north, permanently frozen soil or permafrost presents unique foundation problems. Frozen soil as such would normally present excellent foundation support, since the ice forms a strong bond between the soil particles. The difficulties are caused by the fact that the upper layer of permafrost thaws during the brief summer period. This upper layer, subjected as it is to repeated freezing and thawing with the seasons, is called the active layer; it generally increases in thickness where structures, roadways, and runways promote greater heat transmission to the soil. They do this in two ways: by absorbing more radiational heat (since they are usually darker than the natural surface) and by removing the natural insulation layer from the ground surface. Heated structures transmit additional heat to the underlying soil.

The thawed layer is very wet from melted snow and ice, and the water cannot drain off through the underlying frozen soil. It is consequently a poor foundation material. As a result of this condition a number of special foundation methods have been developed for permafrost construction. Piles are frequently resorted to, but care must be taken to ensure that repeated thawing and freezing in the active layer does not jack up the piles (and supported building). The piles are steam-jetted into the frozen soil, and tapered piles are positioned with the larger end down to reduce uplift from frost heave in the active zone. Generally the pile is made to penetrate the permanently frozen ground to a depth equal to twice the thickness of the active layer. Insulation in the form of an air space or gravel blanket is provided between the underside of heated buildings and the frozen soil.

Topography. The topography of a building site can also have its influence on the choice and design of foundations, especially in hilly terrain. When construction is contemplated on or near a slope, whether it

FIG. 8–14   *Creep on hillside.*

is a fill, cut, or natural slope, the continued stability of the site and the effect of the construction on the slope stability must be investigated. The question of the stability of slopes has been discussed in Chapter 6, and obviously construction can only be considered on stable slopes and hillsides. Except for very large structures, the weight of a structure is usually very small compared to the earth mass of a hillside, and the structural weight will not materially affect the gross stability. The soil cover and the upper layer of weathered bedrock may seem excellent support for a structure on level land, but on a slope these layers may be subject to downhill migration (Fig. 8–14). This phenomenon can often be discovered from the crooked growth of trees, leaning power poles, cracks in paving, and in severe cases even cracks in the soil. The creeping material cannot be used for foundation support, and the foundations must be carried down below this layer.

The downhill movement of the creep zone, slow as it may be, can result in significant lateral thrusts against deep foundations, particularly against single piers surrounded by soil. The magnitude of this lateral force cannot be determined very easily, because under such slow deformation the deformation properties of the soil are similar to those of a viscous material. A conservative approach would be to assume passive pressure of the creeping mass against the foundation. However, in view of the slow deformation, plastic flow will develop, and one may consider the creep mass as a fluid with a unit weight equivalent to that of the soil. Figure 8–15 shows more examples where the topography has affected the foundation design.

Adjoining Structures.   The pressure bulb of stress distribution as shown in Fig. 8–3 clearly indicates that even when the foundations of a structure are kept inside the property lines, the soil is stressed by the foundation well beyond the site boundaries. If a structure is in existence on the adjoining site, the foundations for the new building will normally be taken below those of the existing building in order to avoid additional stress on the foundations, the basement walls, and the soil immediately below the foundations of the existing building. Notwithstanding this precaution, the effect of additional stress on the soil below the existing structure must be investigated. The settlement calculations for the new building may have to be modified for the conditions at the property line where the soil already is stressed higher than elsewhere.

An added complication is the difficulty of constructing the new foundations alongside the existing structure without disturbing the foundations of the latter. If only a few individual footings are to be located on the property line, it is usually possible to limit disturbance by using drilled piles or piers. In many instances, however, the new building will have a basement which is deeper than the foundations of the existing structure, and this will involve a continuous excavation along the property line. The law and the building codes in the United States and in many other countries set a limit of one basement story (10 to 12 ft) for the depth of disturbance that must be anticipated in the future for any structure close to the property line. For example, if someone owns a building with foundations near the surface, he must anticipate the necessity of lowering those foundations near the property line to 10 ft below the street level whenever his neighbor wants to build a structure with foundations or a basement at that depth.

If, however, the new building requires deeper excavation, its builder

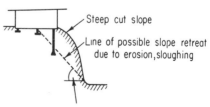

Steep cut slope

Line of possible slope retreat due to erosion, sloughing

Angle determined from orientation of joints, beds, and from performance of similar slopes

FIG. 8-15 *Examples of topography affecting foundation design.*

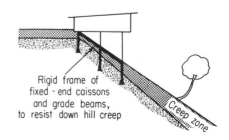

Rigid frame of fixed - end caissons and grade beams, to resist down hill creep

Creep zone

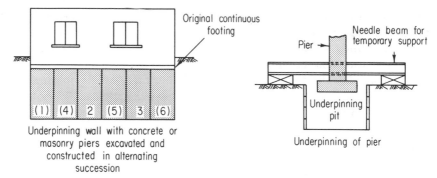

FIG. 8–16   *Typical examples of underpinning.*

must take measures to protect his neighbor's foundations. The most common method of doing this is by underpinning, i.e., by providing new support for the old foundations at a lower level. This can be a difficult operation which usually requires an experienced contractor specializing in this type of work. The foundations are actually undermined for narrow sections at a time while the new supports are being constructed. Another method which appears less complicated but which is in fact more difficult to achieve and more risky is to brace the excavation in such a manner that no lateral displacement of soil below the existing foundations will occur. This involves tight, close, unyielding sheeting or lagging and carefully designed and installed bracing.

The soil engineer should be aware of complications such as these whenever foundations are planned close to existing structures, and both the foundations of existing adjoining buildings and the possibility of future construction on adjoining property must be investigated. Several examples of foundation construction on the property line and underpinning of adjoining buildings are shown in Fig. 8–16.

PROBLEMS

8-1   A three-story building is to be constructed on a sand beach. Groundwater rises to a maximum of 8 ft below grade. The beach sand has the following properties: $\gamma = 110$ pcf, $\phi = 32°$. The maximum column load will be 150 kips. What would be a practical footing size and depth, using a factor of safety of 3.0? (Settlements are not to be considered here.)

Answer   Equation 6–12, adapted to square footings, gives

$$p = 0.8 N_\gamma \gamma b + N_q \gamma h + 1.3 N_c c$$

In sand, $c = 0$. Table 6–9 gives, for $\phi = 32°$, $N_\gamma = 30$ and $N_q = 25$. Ultimate bearing capacity for a footing with a width of $2b$ ft at a depth of

h ft below grade is $p = 5,280b + 2,750h$ psf. Applying factor of safety of 3.0 gives $p_{\text{allow}} = 1,760b + 917h$. It is not advisable, if it can be avoided, to take a foundation in sand to within less than about 2 ft above the water table. The allowable bearing capacity for a maximum practicable h equal to 6 ft is $p_{\text{allow}} = 1,760b + 5,512$. The first part of this answer should be reduced to take into account that part of the soil beneath the footing is submerged. However, the required footing width in this case is such that only a negligible amount of submerged sand is beneath the potential slip surface. Total load is 150 kips, or $150,000 = (1,760b + 5,512) \times (2b)^2$. This gives approximately $b = 2$ ft, so that footing width is 4 ft and $p_{\text{allow}} = 9,400$ psf. However, excavation in sand is costly, since it requires shoring or flat excavation slopes. It will be more economical to construct a shallower, larger footing. Near a beach there is generally little danger of frost penetration, and consequently the footing depth could be as little as 2 ft. For $h = 2$ ft, the footing width becomes roughly 5 ft. The cost of the added 12 in. in footing dimension is well outweighed by the savings in excavation and construction over those for a deeper footing.

8-2 The Uniform Building Code contains a table, No. 28-B, for allowable soil pressures. Two of these allowable soil pressures given for a 1-ft-wide strip footing founded at 2 ft below the surface are as follows: for compact, coarse sand, $p_{\text{allow}} = 1,800$ psf; for soft clay, $p_{\text{allow}} = 1,000$ psf. (Soft clay is usually defined as having a cohesive strength of from 250 to 500 psf.) If the compact, coarse sand has a friction angle $\phi = 38°$ and a unit weight $\gamma = 130$ pcf, what is the factor of safety (F.S.)? What is the factor of safety for a soft, saturated clay with a cohesive strength $c = 250$ psf?

Answer For $\phi = 38°$, $N_q = 50$, $N_\gamma = 75$, $b = 0.5$ ft, $h = 2$ ft. This gives an ultimate bearing capacity $p = 17,900$ psf, and

$$\text{F.S.} = \frac{17,900}{1,800} = 9.9$$

If saturated clay, $\phi = 0$ (initially) and $N_c = 5.7$. Thus,

$$p = 5.7 \times c = 1,420 \text{ psf} \quad \text{and} \quad \text{F.S.} = \frac{1,420}{1,000} = 1.4$$

In the sand the safety factor is unnecessarily high and in the clay it is dangerously low and may result in considerable settlement and plastic deformations.

8-3 The two principal soil layers at a piece of reclaimed waterfront land are a 5-ft layer of dense, hydraulically placed sand ($\phi = 32°$, $\gamma = 120$ pcf), underlain by medium stiff clay ($c = 600$ psf, $\gamma = 110$ pcf). Footings may be founded a minimum of 2 ft below existing grade. Groundwater is more than 10 ft below the surface. What will be the maximum allowable bearing pressure (F.S. = 3.0) for (a) a strip footing with a line load of 5 kips/ft;

(b) a strip footing with a line load of 25 kips/ft; (c) a square footing with a load of 100 kips? Settlements are not considered in this problem.

Answer   (a) The ultimate bearing capacity of a strip footing on the sand is $p = 30\gamma b + 25\gamma h$ psf. With an F.S. of 3.0 and $h = 2.0$ ft this gives $p_{allow} = 1,200b + 2,000$. For a 5-kips/ft load this gives $b = 0.8$ ft and $p_{allow} = 5,000/(2 \times 0.8) = 3,125$ psf. A check must be made on the pressure increase in the clay. At the surface of the clay the ultimate bearing capacity $p = 5.7c + \gamma h = 3,420 + 600 = 4,020$ psf. For an F.S. = 3.0, $p_{allow} = 1,340$ psf. Use Kögler's simplified method (Fig. 8–10) to determine the pressure increase on the clay. The width of the strip load on the clay is $(1.6 + 2 \times 3 \tan 55°) = 10.2$ ft. Maximum pressure increase σ_z on clay is 850 psf. (b) For a 25 kips/ft load on the sand, b becomes 2.5 ft and $p_{allow} = 5,000$ psf. However, the pressure increase on the clay would then be 2,700 psf, greater than the maximum allowable of 1,340 psf. The latter value governs and the footing width must be increased to 14.4 ft, for a net pressure on the sand of 1,720 psf. (c) The allowable bearing pressure on the sand is, for a square footing, $p_{allow} = 960b + 2,000$. This gives, for a 100-kip load, $b = 2.4$ ft and $p_{allow} = 4,350$ psf. To determine the pressure increase on the clay, the Kögler diagram may be extended for both horizontal axes. The area under added stress at the surface of the clay layer will be 13.4 ft square, and $\sigma_z = 1,200$ psf. The allowable pressure on the clay for a square load is 1,680 psf.

8-4   The subsoil at the site for a 50-ft-diam storage tank consists of 30 ft of soft, organic soil followed by 15 ft of dense sand underlain in turn by a thick layer of firm ($c = 1,000$ psf) but quite compressible clay. The tank, with a full weight of 6,000 kips, is founded on a strong, rigid, reinforced-concrete mat supported by end-bearing piles driven 5 ft into the sand. The pile spacing is such that each pile supports an equal area of pile cap. Settlement calculations show 5 in. of settlement for the tank and its pile foundation. This is considered acceptable. Do you consider this a good foundation design under the circumstances?

Answer   No. The clay will not be overstressed below the sand blanket to the point of failure, but the calculated settlement will be about twice as much under the center of the pile group as under the perimeter. Because of the rigidity of the mat, the perimeter piles will want to move down with the central piles, and eventually the perimeter piles will carry substantially more than their original share of the load. The pile spacing should be modified to provide closer pile spacing at the perimeter.

8-5   The soil at the location of a bridge abutment consists of 20 ft of clay underlain by dense sand. Embankment fill up to 15 ft high will be placed on both sides of the abutment. Settlement under this fill weight is expected to be 5 in. prior to embankment construction and 2 in. after completion of the bridge. The abutment will be founded on cylindrical, 12-in.-diam closed-end pipe piles with the pile cap near existing grade and driven

3 ft into the dense sand. Groundwater is 10 ft below existing grade. The soil properties are as follows: fill, $\gamma = 125$ pcf; clay, $\gamma = 110$ pcf, $c = 400$ psf; sand, $\gamma = 120$ pcf, $\phi = 36°$. What is the allowable bearing capacity of the piles?

Answer   End-bearing capacity for a pile with so little penetration in the sand is as for circular footing: $p = 0.6N_\gamma(120 - 62.4)b + N_q q$. For $\phi = 36°$ this gives $p = 180,000$ psf. Allowable load on pile tip (F.S. = 3.0) is $(180,000/3) \times \frac{1}{4}\pi = 47$ kips. "Negative friction" or "downdrag" on pile during settlement of clay is $0.9 \times \pi \times 1.0 \times 0.400 \times 20 = 22.5$ kips. Remaining allowable design load is $47 - 22.5 = 24.5$ kips per pile.

8-6   The ratio of available allowable bearing capacity to the capacity lost to downdrag in the above case was 47/22.5. What is this ratio for (a) an 18-in.-diam cylindrical pile, (b) a tapered pile with a 14-in.-diam butt and an 8-in. tip, (c) a 12-in.-square pile with a tip enlarged to 30 in. square? Ignore minor influences on the perimeter shear such as that of the taper and of the extra disturbance from driving the enlarged foot.

$$\text{Answer} \quad (a) \; \frac{105.8}{33.8}, \; (b) \; \frac{21}{21}, \; (c) \; \frac{379}{28.5}$$

REFERENCES

1. F. A. L. SMITH, Pile Driving Analysis by the Wave Equation, *J. Soil Mech. Found. Div. Am. Soc. Civil Engrs.*, vol. 86, August, 1960.
2. N. M. NEWMARK, Influence Charts for Computation of Stresses in Elastic Soils, *University of Illinois Bull.* 38, 1942.
3. C. W. DUNHAM, "Foundations of Structures," 2d ed., McGraw-Hill Book Company, New York, 1962.
4. LEONARDO ZEEVAERT, Foundation Problems Related to Ground Surface Subsidence in Mexico City, *ASTM, STP* 322, 1963.
5. A. W. SKEMPTON and D. H. MacDONALD, The Allowable Settlements of Buildings, *Proc. Inst. Civil Engs. (London)*, vol. 5, no. 3, December, 1956.
6. A. DÜCKER, Is There a Dividing Line Between Non-frostsusceptible and Frostsusceptible Soils? *Tech. Transl.* 722, National Research Council, Ottawa, Canada, 1958.

9 | Retaining Structures

9-1 DEFINITION

An earth-retaining structure is necessary wherever a sharp drop in the level of the ground surface is required and where the soil is incapable of supporting itself in a vertical or near-vertical slope. The principles of slope stability were presented in Chapter 6, and it was pointed out that temporary vertical cuts can be made safely in many cohesive soils, but that eventually such cuts will require support as a result of long-term processes at work in clay soil. Consequently, virtually all soils can be considered to require support for permanent, near-vertical slopes. Retaining structures are very much in evidence along the waterfront, where they often serve a dual purpose of shoreline protection and mooring facility. Building basements and highway fills are frequently bounded by soil-retaining structures. Other structures which must be designed to retain soil pressures are culverts and tunnels, which are subjected to both lateral and vertical pressures.

In Chapter 6, the theory of active and passive earth pressure and of stability of slopes and retaining structures was presented. The purpose of the present chapter is to describe a number of the most common forms of retaining structures, the method of design so far as it is concerned with the soil properties, and the method of construction. The factor of safety to be applied in the design of the various types of structure will be discussed separately for each type. This is done because the safety factor of a retaining structure must be established on the basis of a large number of conditions: the consistency of the soil properties, the accuracy of the design assumptions for earth-pressure distribution, the intermediate stages of loading during construction, the purpose of the structure, and the consequences of failure.

It was explained in Chapter 6 how the minimum or active lateral pressure on a retaining structure can be visualized as developing when the structure moves or yields laterally in the direction of the pressure. From this principle it follows that a structure incapable of movement in that direction will have to support a higher pressure. As a rule, the movement required for an active condition to develop is very small, of the order of $\frac{1}{4}$ to $\frac{1}{2}$ percent of the wall height at the top, and even a very rigid structure without restraint near the top is capable of some movement by rotating with its footing. If the upper part of the structure is braced, however, as with shoring or with basement ceiling beams, no movement can develop except by sliding of the foundations, and the latter is unlikely to occur in a properly designed structure. For this reason a distinction has been made between structures with and without lateral restraint above the foundations.

The last section of this chapter describes special marine structures, conduits, and tunnels.

In special cases, backfill material such as ashes of lighter than usual unit weight is employed to reduce the load on a retaining wall. In other circumstances, backfills or surcharges of high unit weight may have to be designed for, when, for example, a retaining wall or wharf for an iron ore storage area is required.

9-2 STRUCTURES WITHOUT LATERAL RESTRAINT ABOVE THE FOUNDATIONS

All structures falling in this category are of a permanent nature (i.e., not used for temporary support of cuts or excavations), and in virtually all cases they will at least partially retain backfill material placed after completion of the structure. This has an advantage in that the type of backfill can be selected in general to be a noncohesive soil. The pressure on a wall which for any reason will be able to yield somewhat under the soil pressure, as mentioned above, may be expressed as

$$p_a = K_a \gamma z \tag{9-1}$$

where K_a is directly related to the angle of internal friction ϕ and z is the depth below a level backfill surface. For clay backfill against this type of wall the pressure will vary with time. Saturation may lead to an ultimate value of K_a of 1.0, so that the pressure will correspond to that of a liquid with the soil unit weight. Consequently, sand backfill not only has the advantage of generating a lower active pressure than cohesive soil; it also results in a more predictable pressure. Furthermore, if the sand is drained by an underdrain or weep holes in the wall (as it always is in practice), the accumulation of water behind the wall, with its resultant high lateral pressure, will be prevented.

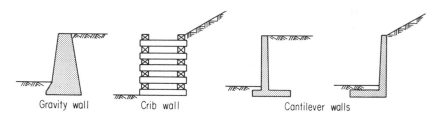

Gravity wall   Crib wall   Cantilever walls

FIG. 9–1   *Typical retaining walls on spread footings.*

Some typical retaining walls of the type discussed here are shown in Fig. 9–1. The first two walls shown are gravity walls, which depend on their own weight for stability against overturning. The last two walls shown are cantilever walls of reinforced concrete or masonry. The second wall is a so-called crib-type or bin-type wall, which consists of stacked reinforced-concrete or steel members filled with sand. In the design of the retaining structure it is first of all necessary to estimate the magnitude and the direction of action of the earth pressure on the wall.

The active earth pressure from a cohesionless soil on retaining walls of the types shown is related to the design variables β and ω indicated in Fig. 9–2, by the soil properties γ and ϕ and by the friction angle between the wall and the soil δ. The influence of δ on the magnitude of the active pressure is small and can usually be ignored. It is evident, however, that δ has a considerable influence on the stability of the wall against overturning. As mentioned in Chapter 6, δ usually can be taken as $\frac{2}{3}\phi$ for walls of concrete or similar material. Elaborate tables are in use for finding the coefficient of active earth pressure for different values of ϕ, β, and ω.[1] Table 9–1 gives some typical values for K_a. Cohesive soils behind permanent walls must be treated as noncohesive for long-time performance. In extreme cases K_a may approach 1.0 in clays, but if saturation in the future can be avoided, a value of 0.5 may be assumed safely.

If the soil surface behind a wall has an irregular shape, or if the surface is surcharged with a uniformly distributed pressure, the active earth pressure can be determined graphically by using the method of Coulomb and Culmann's graphical solution as given in Fig. 6–13. The effect of incidental surcharges such as line or point loads on the surface behind the wall must be analyzed differently, and this is commonly done by using the

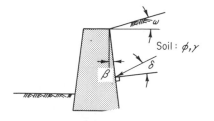

Soil: ϕ, γ

FIG. 9–2   *Retaining-wall parameters.*

TABLE 9-1

β	$\phi = 30°$	$\phi = 35°$	$\phi = 40°$	ω
0°	0.33	0.27	0.22	0°
0°	0.46	0.35	0.27	20°
10°	0.40	0.34	0.28	0°
10°	0.57	0.46	0.36	20°

theory of linear elasticity for stresses in a semi-infinite medium such as was used for determining vertical stresses below footings in Chapter 8. In the case of horizontal stresses resulting from vertical loads, the Poisson's ratio μ of the soil must be given an assumed value. Frequently a conservative value of 0.5, which represents an incompressible medium and is valid for plastic clays, is given, but a value of 0.4 or 0.3 is more appropriate for sands. The horizontal stress in the soil at depth z and a horizontal distance x from a point of application at a point load P (Fig. 9–3) can be expressed as

$$\sigma_x = \frac{P}{2\pi}\left[\frac{3x^2z}{R^5} - \frac{1 - 2\mu}{R(R + z)}\right] \tag{9-2}$$

where

$$R = \sqrt{x^2 + z^2}$$

In this method the stresses are assumed to act on the wall as if it were a vertical plane in a semi-infinite medium at distance x from the load. The effect of a line load can be determined by integrating the same equation. As in the case of vertical stresses, Newmark has prepared influence charts for horizontal stresses acting on vertical planes.

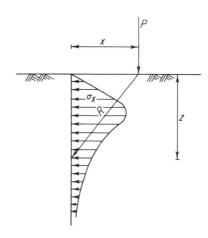

Fig. 9–3 *Horizontal stress increase produced by a point load P.*

There are some cases where a pressure calculation different from the calculations presented above is justified. If, for instance, a retaining wall is built to support a relatively dry clay which by its location will be permanently protected against an increase in moisture content, it is quite acceptable to make use of the relatively high shear strength a dry soil of this type may have. A relatively dry or partially saturated clay has a high shear strength which is partly caused by negative pore pressure from capillary suction (see Chapter 3), and if this negative pressure can be expected to remain for the lifetime of the structure, it would be unnecessarily conservative to ignore this strength. Except for great heights, this usually means that the vertical soil surface can support itself without a wall, and the wall serves only as an additional safeguard. Other materials in which lower earth pressures may be employed are cemented soils and weathered bedrock formations, which need some support against deterioration but which may have substantial inherent shear strength. In the latter case, the possibility of joint- or bedding-plane failures must be investigated.

After the magnitude and the direction of the earth pressure on the retaining structure are established, the design and the method of constructing the foundation can be considered. Because of the differences in design, retaining walls on spread footings and on piles will be discussed separately.

Foundation on Spread Footings.   Walls founded on spread footings normally look like those shown in Fig. 9–1. With the earth pressure

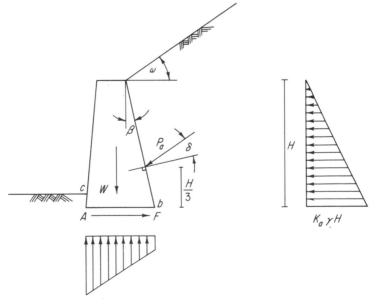

FIG. 9–4   *Gravity retaining wall in sand.*

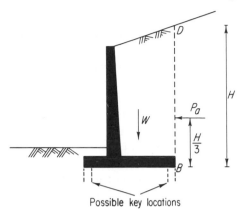

FIG. 9-5   *Forces on a cantilever type retaining wall.*

known in magnitude and direction, the wall must be designed in such a manner as to prevent (1) the footing pressure from exceeding the maximum allowable pressure from either a stability or a settlement point of view, (2) overturning of the wall, (3) sliding of the wall, and (4) structural failure of the wall.

Figure 9-4 shows a gravity-type retaining wall on a spread footing. For the soil under the entire footing to be in compression, the resultant of the weight W and the pressure P_a must intersect the base Ab within one-sixth of its width from the center of the base; i.e., the resultant must lie within the middle third of the base. This also provides a safeguard against overturning. Overturning can be checked by knowing the magnitude of the overturning and restraining moments about point A. Sliding along the base is rarely a problem for this type of wall; for concrete poured against a rough, undisturbed soil base the friction coefficient along the base may be taken as equivalent to that of the soil. The effect of passive resistance against the toe Ac is usually not included in the calculations because of uncertainty about the continued presence of the soil cover at the toe. If the depth of foundation is substantial (frost protection), part of it may be used for passive resistance, provided the toe is poured against undisturbed soil.

In Fig. 9-5 a cantilever-type wall is shown; it has substantially the same system of forces as the gravity wall. Because it is unlikely that friction at the stem of the wall would still have appreciable effect at the line BD, δ can best be taken as zero, with P_a acting horizontally. With this wall, the possibility exists that not enough frictional resistance will develop along the base, in which case a key can be used to generate passive resistance. At first sight, the best location for a key would seem to be at the heel, as shown in Fig. 9-5. The effect of a key at the heel is to lower the potential plane of sliding, which results in a greater shearing resistance along it. On the other hand, a key at the toe would add passive resistance of the same magnitude as the key at the heel, in addition to frictional resistance along the base. In practice, therefore, the key location may be selected where convenient.

Factors of safety normally used in designing these types of retaining walls are different for different possibilities of failure. The factor of safety against overturning is determined as the ratio of the moments about point A of the two forces P_a (overturning) and W (restoring); it is normally specified as to be at least 2.0. A factor of safety of 1.5 against sliding is frequently used; more often, however, a maximum allowable friction coefficient of 0.4 or 0.5 is specified, which results in a safety factor of about 1.4 for most soils. Unless considerable uncertainty exists about the method of backfilling, the adequacy of water drainage from behind the wall, or surcharges on the backfill, no safety factor is applied to the actual lateral pressure; adequate safety is available if the structural members of the wall are designed along conventional lines.

Foundation on Piles.   If the soil immediately below a retaining wall cannot safely support the vertical and lateral loads, a foundation on piles must be used. Because piles are expensive and because pile foundations in most cases can be designed to support inclined loads or even upward loads, the gravity-type wall is very seldom used when a pile foundation is required, and the cantilever wall is the most common type. The weak soil (usually clay) found to be incapable of supporting the wall may also be stressed close to failure by the weight of the backfill behind the wall, and a failure of the type shown in Fig. 9–6 must always be investigated. Even if no failure appears imminent, plastic deformation of the soil below the backfill may cause lateral stress against the piles. It is important, therefore, that the soft soil below the fill not be stressed too close to its yield strength by the backfill or surcharge. The shear strength of the clay can be improved by preloading and by using sand drains, as will be discussed in Chapter 10.

Another effect of the application of a high fill load on the clay behind the wall is consolidation of the clay, which will cling to the piles as it settles and reverse the direction of the friction on the piles. This *negative friction* can result in a very considerable increase in loading on the piles. The loading due to negative friction can conservatively be estimated to be

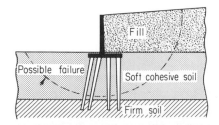

Fig. 9–6   *Cantilever-type retaining wall on pile foundation.*

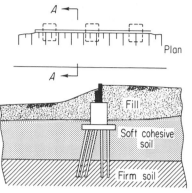

FIG. 9-7 *Spill-through abutment.*

equal to the cohesive strength of the clay times the perimeter area of the piles. In some cases, the combined effect of lateral pressures from plastic deformation of the clay and additional vertical load due to negative friction is such that a continuous abutment of the type shown in Fig. 9-6 is not feasible. Instead, a *spill-through abutment* may be used; it is basically an abutment divided into three or more sections with space between sections below a certain level (Fig. 9-7). This reduces in particular the lateral forces due to yielding of the clay under the abutment fill, whereas negative friction is limited mainly to the perimeter piles in each group. The amount of settlement due to consolidation of the clay must also be calculated.

If the soil near the surface is reasonably firm and granular (piles may still be needed for scour protection), vertical piles can support part of the lateral load. A theoretical determination of the lateral pressure distribution, against a pile which is pushed horizontally into the soil is quite difficult, since it involves the deformation properties of both the pile and the soil. As a rule, a lateral force of 5,000 lb for a 12-in.-diam pile is considered allowable if the piles are driven into medium dense sandy soil or into medium stiff clay. In soft clays, the long-time deformation of the soil around the pile would be too great, and the lateral component must be taken by angled or batter piles. For temporary (wind or seismic) lateral loading, the resistance in clays can be expected to be higher; values up to 10 kips per pile are quite common.

9-3 STRUCTURES WITH LATERAL SUPPORT ABOVE THE FOUNDATIONS

Rigid Permanent Structures. It was seen in Chapter 6 that a condition of active earth pressure or a condition of minimum lateral stress in relation to the vertical stress can develop only upon movement of the retaining structure in the direction of the lateral stress. Conversely, the condition of maximum or passive pressure can develop only when the wall is moved against the direction of the lateral pressure. What kind of pressure must be expected when the wall is completely rigid and essentially cannot move in either direction? A stress condition somewhat in between the active case of Fig. 6–3 and the passive case of Fig. 6–6 will develop. The pressure will depend, among other things, on the method of placing the backfill.

Since the passive pressure usually is several times greater than γz, the vertical stress, it is very important for design purposes to be able to approximate whether the actual condition in case of a rigid, immovable wall is closer to the active or to the passive condition. It is logical that without movement of the wall against the soil the pressure on the wall cannot be greater than the vertical stress; even if the soil were in the state of a highly viscous fluid the lateral pressure would only just equal the vertical stress. This stress state is called a neutral or at-rest condition, and the lateral pressure on the wall is called a neutral pressure, $K_n \gamma z$. It will develop against basement walls which are held in place by floors at the top and the bottom and against rigid frame bridge abutments, culverts, and the like. K_n will have a value between that of K_a and 1.0, depending mainly on the degree of compaction and the type of backfill used or the condition of the natural soil retained.

If a sandy backfill is placed against a rigid wall without an effort to compact it, the resulting pressure will not differ much from that of the active condition. Field and laboratory experiments have shown K_n to be of the order of 0.4 to 0.5, whereas laboratory experiments with highly compacted sandy backfills have produced K_n values up to 1.0. It is doubtful if such a value could ever be obtained in the field. For clay backfills, compacted at a low moisture content, the initial pressure will be quite similar to that of compacted sand, but future saturation may increase K_n to close to 1.0. For this and other reasons clay backfills are not used very often.

It has become common practice to design rigid retaining walls of the kind discussed here on the basis of a lateral pressure linearly increasing with depth at the rate of 30 psf/ft, or somewhat higher if the soil slopes upward from the wall. This value can be found in many building codes and design handbooks, usually accompanied by recommendations for compacting the backfill. Since the unit weight of most compacted soils will be about 110 to 120 pcf and the coefficient of neutral earth pressure of a compacted sand, as mentioned, can be expected to be 0.5 or even more, the conclusion must be drawn that many basement walls and similar structures are designed for pressures which are very much below those actually occurring. Nonetheless, failures of such walls are extremely rare, and those which have taken place usually could be ascribed to surcharges resulting from compaction or construction equipment. This situation arises for several reasons. In the first place, in the design, friction against the wall is neglected ($\delta = 0$) and the resulting deflection of the lateral force from the horizontal is not taken into account. Second, the structure usually is designed as a series of one-foot-wide individual strips, whereas in most cases it is a rigid membrane often buttressed at short distances by basement walls. Third, overstressing of a wall will cause some deflection which immediately results in a reduction of the pressure toward the active value. Finally, the factor of safety of the actual structure may be quite high.

Occasionally, a retaining structure of the type just described can be constructed by pouring it directly against a natural soil which is able to support itself temporarily in short sections in a vertical cut. In that case, the pressures and pressure distribution will be different from those developed by a backfill and will be more like those in braced cuts, which are described in the following section.

Braced Cuts. The difference between braced cuts and structures designed to provide rigid support only after their completion to full depth is that the bracing and shoring in a cut is placed successively as the natural soil is removed from the excavation and supports the remaining soil in the in-place condition. Instead of carrying load from a backfill placed against the wall from the bottom upward, the bracing will begin to take the soil's pressure as the cut proceeds downward. In Fig. 9–8 three successive stages of excavating and bracing of a cut are shown with the resulting lateral pressure distribution on the shoring for each stage. In Fig. 9–8a, excavation has proceeded to the level of the first brace. The soil in this example has some cohesion or temporary bond, so that excavation to depth h can be made safely without support. Shoring and a brace are installed to this depth. When sheet piling is used, the piles are driven first and are braced as excavation proceeds. The bracing is given support by the opposite side of the excavation or by deadmen or other forms of

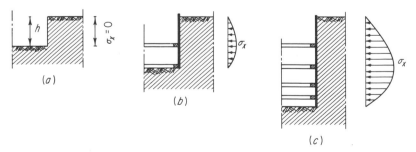

FIG. 9–8   *Development of pressure against shoring of a braced cut in clay.*

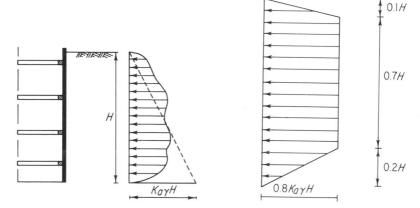

Fig. 9–9   *Pressure on shoring of a braced cut in sand.*

Fig. 9–10   *Tschebotarioff's trapezoida pressure diagram for braced cuts in sand*

anchoring that provide as nearly as possible unyielding restraint to the shoring.

The shoring itself is usually some form of sheeting or lagging which may be comparatively flexible between major supports. This has the result that, between braces, the retaining structure is usually capable of some movement away from the soil, but at the points of bracing little or no movement is possible. Consequently, the total lateral force per lineal foot of excavation will be somewhat higher than that created by an active condition, and the pressure distribution will no longer be triangular. Strictly speaking, it varies both along and up and down the wall.

Because bracing is installed as the excavation goes down, deflection of the restrained upper part of the support will be very small, and the final pressure on the upper part will be considerably higher than the active pressure. The pressure diagram may look similar to that shown in Fig. 9–9. For design purposes, the diagram is commonly assumed to be a trapezoid in sandy soil; the trapezoidal distribution of Fig. 9–10 was proposed by Tschebotarioff.[2] It can be seen that an analysis on the basis of a triangular distribution would give loads on the upper struts well below those actually occurring, even if a conservative value for the active earth pressure coefficient has been assumed. Failure of the upper strut of braced cuts has not been uncommon in practice.

For braced cuts in clay, the pressure diagram of Fig. 6–4 requires modifications similar to those required in sandy soil. For a cut in a nearly saturated clay with a cohesive strength c, the sides will not require support until the height of the cut exceeds the approximate value $H = 4c/\gamma$ (Chapter 6). In saturated, plastic clays this will be the case for only a short period; the resulting strain in the clay will cause increasing deformations at the cut face, and a bulge-type failure could develop. To arrest

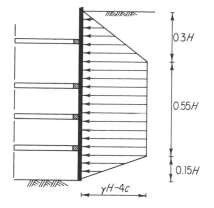

FIG. 9-11   *Peck's lateral earth-pressure diagram for braced cuts in plastic clay.*

this gradual squeezing-in of the excavation sides, a resisting pressure well in excess of the active pressure determined from the laboratory shear properties will be required. This is particularly important if no deformation can be allowed during the period the excavation is open.

Peck[3] has suggested a lateral earth pressure diagram for the design of braced cuts in plastic clays, based on the unconfined compressive strength of the clay (Fig. 9-11). For these clays the unconfined compressive strength q_u may be taken as $2c$. This method will provide quite heavy bracing near the top, which is where bracing is known to have failed when designed on the assumption of a triangular distribution. The condition for the successful use of Peck's method is that the unconfined compression tests must be stopped at 5 percent strain of the sample and that the strength at that strain is considered the strength at failure.

Tschebotarioff[2] has introduced the neutral earth pressure ratio method for clays classified as stiff, medium stiff, and soft (Fig. 9-12). This represents a return to the triangular distribution, using a coefficient of earth pressure at rest K_n of 0.5 and allowing a reduction of the stresses on the lower part of the shoring, depending on the stiffness of the clay. It is felt that for stiff, overconsolidated clays the neutral earth pressure method may result in too low a stress on the upper bracing, and Peck's method would be more suitable. Tschebotarioff suggests that either Peck's method or the neutral earth pressure method can be selected, depending on which gives the *least* favorable results. In general, the neutral earth pressure

FIG. 9-12   *Tschebotarioff's   neutral-earth-pressure ratio method for braced cuts in plastic clay.*

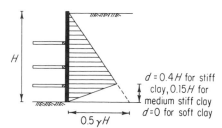

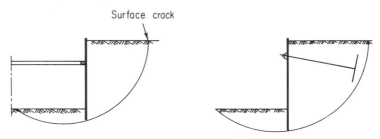

Surface crack

FIG. 9-13   *Possible failures of braced shoring with earth support below the excavation bottom.*

ratio method is considerably more conservative and can be used safely for preliminary design purposes based on soil classification. Peck's method usually provides a more economical design which can be used safely for clays whose yield strength is considerably higher than the strength at 5 percent strain.

In cases where the lower part of the shoring receives support directly from the soil below the bottom of the excavation (Fig. 9-13), the possibility of a rotational failure of the type described in Chapter 6 must be investigated, particularly in wide excavations. As can be seen, this is particularly important in the case of anchored shoring, where the anchor may be within the potential sliding mass. The situation can be improved by extending the shoring to a greater depth into the soil and by increasing the length of the tie bars to the anchors.

The effect of a surcharge on the soil to be retained in a braced cut will also be somewhat different from that on a retaining wall which can move at the top. In addition, the effect of a surcharge already in place will be different from the effect of one which will be applied after the cut and the bracing have been completed. If a surcharge is already in place when the cut is made, elastic deformation has taken place in the soil. Subsequent excavation some distance away from the surcharge will remove some of the elastically strained soil, causing a redistribution of stresses, but unless this redistribution causes movement or plastic deformation in the direction of the shoring, there will be no increased pressure on the shoring. If the possibility of movement exists (Fig. 9-14), the shoring must be

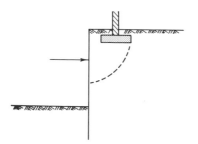

FIG. 9-14   *Surcharge in place before excavation. Restraining force must provide factor of safety against failure of existing foundation.*

designed to provide for additional pressure. This is an important consideration when an excavation is planned near existing structures. In that case the pressure required to stop movement must be calculated, and the bracing must be prestressed to about twice that value immediately upon installation to allow for stress release during the lifetime of the shoring system.

If a surcharge is placed on the soil retained by shoring already in place, the stress on the shoring can be calculated with Eq. (9–2). However, contrary to the concept outlined for walls without lateral restraint at the top, a braced and relatively unyielding support like braced shoring cannot be assumed to be a continuation of the elastic medium. Therefore, the lateral pressure on the bracing of a cut or on any rigid, laterally restrained retaining structure, caused by a surcharge placed behind that structure subsequent to its installation, will be approximately given by considering two surcharge loads symmetrically placed with respect to the retaining structure (Fig. 9–15). One of these loads is fictitious, but its effect is to give a vertical plane at the position of the bracing which undergoes no horizontal deflection. In this case, by superposition, the lateral stresses on the bracing given by Eq. (9–2) are doubled. This will be significant only for rapid load applications; consolidation and plastic flow or creep will tend to compensate for the rigidity of the wall, causing a reduction of the initial stress on the wall.

The simplest method of supporting the sides of an excavation is, of course, to brace against the opposite side, as shown in Fig. 9–16. When this is not possible, the lateral resistance must be obtained from the soil below the excavation bottom or behind the retaining structure. Figure 9–17 shows several common methods of supporting shoring. The first system is the traditional method used in construction. Soldier piles are driven or drilled in place before excavation begins. They may consist of steel

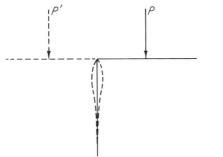

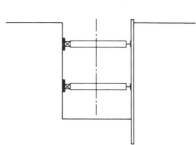

FIG. 9–15 *Fictitious surcharge load, placed symmetrically with respect to rigid, nonyielding wall, would prevent horizontal movement of wall.*

FIG. 9–16 *Open cross bracing (left) and sheet piling with cross bracing (right) in trench.*

H-piles or slotted cast-in-place concrete piles. The spacing of soldier piles is usually less than 8 ft, and timber, steel, or concrete planks or lagging can be fitted between the piles as the excavation proceeds. The piles are straddled by walers, and inclined braces provide the lateral support. There will be a substantial upward component against the waler and the shoring, depending on the steepness of the bracing. The braces are supported by deadmen or a pile frame or sometimes by a horizontal ground beam transferring the load to a brace across the excavation.

The second method shown in Fig. 9–17 is an anchored retaining structure, which has the great advantage of providing an excavation which is free from obstructions. The cost of working inside an excavation which is cluttered up with braces and deadmen can be several times that of working in a free excavation. The anchored type support lends itself best to cases where sheet piling or closely spaced soldier piles are receiving substantial support in passive resistance immediately below the excavation bottom and where this resistance combined with the restraint of one row of anchors is adequate for support. The anchor may be a continuous strip, a rectangular plate or board, an A-frame consisting of two piles, or a drilled-out, bell-shaped plug installed with a special horizontal drill. The latter are becoming quite popular, and several systems have proved their worth. With anchors of this type it is advisable to prestress the anchor to its design load both to check on the adequacy and to avoid deflection of the shoring due to initial yield. The use of anchors in combination with verti-

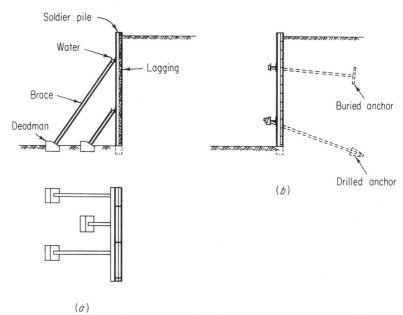

FIG. 9–17   *Braced shoring and anchored shoring.*

FIG. 9–18 *Excavation with cross bracing, soldier piles, and closed lagging.* (Courtesy Skidmore, Owings, and Merrill, San Francisco)

cal sheet piling will be discussed in connection with the flexible bulkheads. When anchors are employed, the ground outside the excavation must, of course, be available for their use. Many variations on the two principal systems of shoring are in use; Figs. 9–18 and 19 show some practical examples of excavation support.

The method of design of deadmen and anchors is based directly on failure mechanisms in the soil such as described in Chapter 3 for the soil below a foundation. Deadmen are constructed in two fashions as shown in Fig. 9–20. The deadman in Fig. 9–20a is used wherever the footings for the permanent structure can be utilized temporarily for support of bracing. The inclined deadman of Fig. 9–20b is more effective, but it is less easy to install. The safe design load for either type of deadman can be calculated with the aid of Meyerhof's[4] semiempirical bearing-capacity coefficients for footings with inclined loading (Fig. 9–21). These coefficients are for use in the equation

$$q = cN_{cq} + \gamma \frac{B}{2} N_{\gamma q} \tag{9-3}$$

where N_{cq} and $N_{\gamma q}$ are bearing-capacity coefficients that are similar to those described in Chapter 5 but are variable not only with the angle of internal friction ϕ but also with the angle of inclination of the load α and the depth-width ratio D/B of the footing or deadman. For anchor walls buried some depth below the surface, the passive resistance can be cal-

FIG. 9–19   *Excavation with anchors, soldier piles, and closed lagging.* (Courtesy Webb & Lipow, Los Angeles)

culated with the equations used before for noncohesive and cohesive soils, resulting in trapezoidal pressure diagrams such as shown in Fig. 9–22. If, as in most cases, the anchor is of limited length in relation to its depth and height, the unit pressure against this anchor block will be considerably higher than against a continuous wall, since the shearing resistance along

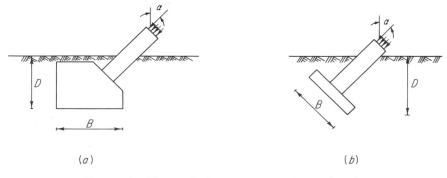

(a) (b)

FIG. 9-20   *Two methods of support for inclined loads.*

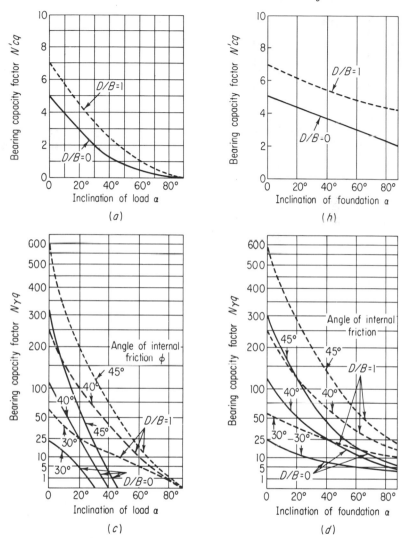

Fig. 9-21 *Meyerhof's bearing-capacity factors for strip footings with inclined load. (a) Horizontal foundation in purely cohesive soil. (b) Inclined foundation in purely cohesive soil. (c) Horizontal foundation in cohesionless soil. (d) Inclined foundation in cohesionless soil.*

four sides of a soil block in front of the anchor must be overcome, rather than only along top and bottom as with a continuous wall.

Experiments have shown that in sandy soils the coefficient of passive earth pressure K_p against a separate anchor block will increase from twice its normal value for $H/h = 1.0$ to as much as three times its normal value for $H/h = 5.0$ or more, where H is the depth of the anchor and h its height

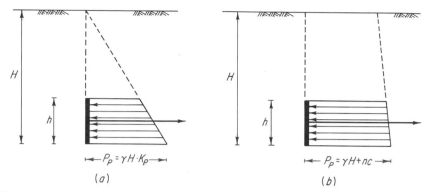

Fɪɢ. 9-22   *Buried anchor blocks in* (a) *sand and* (b) *clay. K_p and n increase with increasing depth-height ratio H/h.*

(Fig. 9–22). In cohesive clays, the independence of strength from normal pressure will eliminate the effect of the surcharge weight γH in the equation for P_p in Fig. 9–22, and the ultimate resistance will be a function of c only once H/h has reached a value of, say, 10 or more. As with deep foundations in clay, the ultimate pressure required to obtain plastic flow around a single block anchor will be of the order of $9c$. For design purposes, a linear increase in passive resistance from $2c$ to $9c$ for a corresponding increase of H/h from 1.0 to 10.0 can be assumed, but the possibility of plastic creep and consolidation at stresses below this ultimate yield stress must be investigated.

Factor of Safety for Braced Cuts.   Shoring of cuts is a temporary phase of construction, and as such it often receives sparse attention. Yet it is possible under difficult circumstances such as an excavation site surrounded by old buildings that the engineering time required to prepare a practical and economical shoring design and plan of excavation is more than that required to design the permanent structure. Adequately designed shoring protects the builder against claims from adjoining property owners, and it protects the lives of the workers in the excavation. The casualties from excavation cave-ins and shoring failures represent the largest percentage of all construction casualties.

The factor of safety must be determined in the first place by the consequences of failure and in the second place by the soil conditions. The temporary nature of the construction would suggest a lower safety factor than for permanent work, but usually the builder more than compensates for this with less workmanlike and less accurate work, and the engineer would be well advised to allow for such common errors as overexcavation, removal of a brace, and surcharging from construction equipment or stockpiles. In most cases it should be a condition of the design that the soil engineer supervises the excavation and shoring work, but unfortunately this is not always so.

If soil conditions and conditions of loading are well known, the factor of safety for passive resistance against deadmen, anchors, or sheet piles should be 2.0. This can be reduced in some cases in granular soils if the design load is applied to the deadmen or anchors by jacks, as it often is to take up the slack in the system. If doubt exists about the effect on the shoring of pile driving, dewatering, or other types of surcharging, the factor of safety against failure of the shoring system should be increased to 3.0. This should also be the case in cohesive soils where plastic creep may develop at the anchors or deadmen.

9-4 ANCHORED BULKHEADS

An anchored bulkhead (Fig. 9–23) consists of vertical, interlocking sheet piles which are embedded in the soil sufficiently deeply to obtain a certain degree of support below the dredge line and which are restrained near the top by tie rods connected to some form of anchor. In most cases, both the anchorage and the flexibility of the sheet piling permit considerable deflection and lateral movement under loading which, as we have seen, results in a reduced pressure on the retaining structure. Anchored bulkheads are frequently referred to as flexible bulkheads, and the design methods discussed in this section apply only to reasonably flexible structures. Anchored retaining structures with rigid support and rigid members should be designed as discussed in the preceding section. The piling is almost always driven first, before the soil on one side is excavated or dredged out to a required depth less than the piling embedment.

In most cases, the height of soil H to be retained and the point of application of the anchor force are determined by site conditions, which leave the type and section of sheet piling and the depth of embedment D (Fig. 9–23) to be determined. In addition to tongue-and-groove timber and reinforced-concrete sheet piling, a very wide choice of interlocking steel sheet piling is available, and practically any required section modulus can be obtained from single or combined sheet-pile sections. Calculation

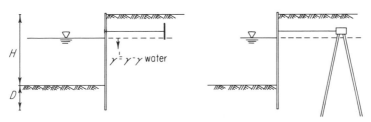

FIG. 9–23 *Examples of anchored flexible bulkheads.*

of the required depth is different for different soil types, and it is based to a very large extent on the results of model tests and full-scale tests, in particular those carried out by Tschebotarioff[5] and Rowe.[6] Only those design methods enjoying frequent practical application will be discussed. Various methods exist because of uncertainty regarding the pressure distribution on a flexible bulkhead.

Bulkheads in Sand. The earliest methods used for flexible bulkhead design both in sand and in clay are known, respectively, as the *free-earth-support* and the *fixed-earth-support* methods. In the first method it is assumed that the embedded part of the sheet piling does not obtain enough fixity to reverse the curvature of the piling and to induce a negative bending moment (Fig. 9–24). The anchor pull must be equal to the difference between the conventionally calculated active and passive pressures, and the depth of embedment is calculated from the sum of the moments around the point of application of the anchor force. A factor of safety of 2.0 is usually applied against failure of soil in front of the bulkhead and against failure of the entire structure. In sands, the method results in a very conservative and expensive design which is justified only when the sheet piling is relatively rigid and when the sand below the dredge line has a low relative density.

The fixed-earth-support method is based on the assumption that the sheet piling acts as a beam with one partially fixed end and a second support at the point of application of the anchor force (Fig. 9–25). The active- and passive-pressure diagrams are the same as for the free-earth-support method, except below point a, where the (reversed) passive resistance is assumed to be concentrated at point a. At point a the sheet piling is assumed to be vertical, and point a is so selected that $D = 1.2D'$. The point of contraflexure c will be below the dredge line, and the method of calculation involves a trial-and-error approach of selecting a depth D. It is

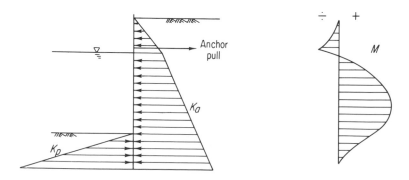

Fig. 9-24 *Pressure diagram and bending moment diagram for free-earth-support method of anchored bulkhead design in sand.*

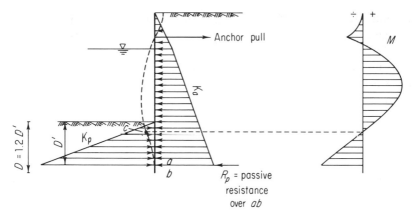

Fig. 9–25   *Pressure diagram and bending moment diagram for fixed-earth-support method of anchored bulkhead design in sand.*

a very slow technique by hand computation and does not offer much advantage over the equivalent-beam method, which will be described next.

Blum[7] simplified the fixed-earth-support method by assuming a hinge at the point of contraflexure c, thereby reducing the problem to one of calculating the equilibrium of two simply supported beams. Blum established a direct relationship between the angle of friction of the soil and the depth of the point of contraflexure below the dredge line. In view of the satisfactory performance of many bulkheads in continental Europe which would have been considered underdesigned according to this equivalent-beam method, the active pressure in Blum's calculations is reduced by 30 percent in several European countries.

Tschebotarioff concluded on the basis of model tests that the point of contraflexure in most sands is very close to the dredge line, and he developed the simplified equivalent-beam method with a hinge at the dredge line. The depth of penetration is set at $0.3(H + D)$, or $0.43H$ (Fig. 9–26). The value for K_a is taken as

$$K_a = \left(1 - \frac{a}{3.5H}\right) \times 0.33 \times 0.9$$

which takes into account the increased pressure at the anchor level at depth a below the surface and a reduction coefficient of 0.9 for the friction against the sheet piling. The method is extremely simple, since it involves only calculating the strength of the upper part of the sheet piling in bending; it may safely be assumed that the bending moment below the dredge line is always smaller than that above it. The determination of the active pressure gives little credit to the soil properties except to the classification as a sand, and in dense, well-graded sands a reduction of the active pres-

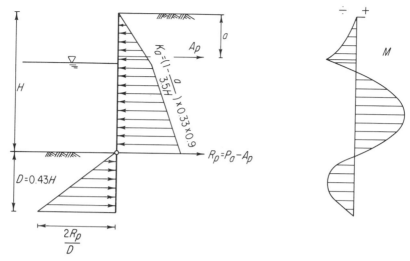

Fig. 9–26   *Tschebotarioff's simplified equivalent-beam method with a hinge at the dredge line for flexible bulkheads in sand.*

sure can be justified. Obviously, a check on the adequacy of the passive resistance against the lower part of the sheet piling will be necessary if soil properties below the dredge line differ from those of the backfill.

Rowe's large-scale model experiments demonstrated the importance of the relative flexibilities of pile wall and soil, and for design purposes he presented his results in the form of diagrams of maximum bending moment ratio versus a flexibility number for the pile. The bending moment ratio is given by the actual maximum bending moment experienced by a pile wall of a given flexibility number divided by the maximum bending moment for the same conditions determined by free-earth-support calculations. The flexibility number depends on the pile material, modulus, and length, and Rowe gives different curves of the bending moment ratio versus flexibility number for various soil types. All the curves are empirically determined from his model tests.

Bulkheads in Clay.   In soft to medium-stiff plastic clays the resistance below the dredge line will frequently be inadequate to provide even partial fixity to sheet piling. Bulkheads are normally used along waterways, where the surficial deposits usually are normally consolidated alluvial soils which cannot provide the resistance required for partial fixity. Consequently, the free-earth-support method, found to be too conservative for sands, is suitable for such clays. Figure 9–27 shows a bulkhead designed in this manner. The upper soil layer behind the bulkhead is usually a sand backfill. The active and passive pressures in the clay are calculated as indicated in Chapter 6, where $q_u = 2c$ is the unconfined compressive strength of an undisturbed sample. As in sand, the minimum depth of penetration required is obtained by taking the moments about the point of application of the anchor force. The depth is then increased to provide a factor of safety of 1.5 or 2.0 against failure.

In stiff clays, partial fixity may be obtained, provided the clay does not lose strength with time. Tschebotarioff has found that contraflexure below the dredge line of sheet piling in stiff clay will take place if the depth of penetration is great enough to maintain the passive pressure in the clay at less than 25 percent of the shear strength. Rowe discovered that the location of the point of contraflexure below the dredge line depends both on the flexibility of the piling and on the compressibility of the soil below the dredge line.

Exact rules for design for bulkheads in silts and other intermediate soils are not available, nor is it possible to present simple rules for bulkhead design in layered systems. The design of flexible bulkheads in complicated soil conditions requires a considerable amount of experience and sound judgment. The premises made for such a design are by necessity somewhat speculative and should be adequately conservative to avoid a failure if the natural soil behavior deviates somewhat from that assumed.

Practically all known failures of flexible bulkheads have been failures of the anchor or tie-rod support or failure of the soil below the dredge line. Failure of the sheet piling in bending is extremely rare, as might be expected, partly because the steel design includes a large margin of safety between the allowable steel stresses and the yield strength and partly because of the reduction of soil stresses which follows yielding. Anchor

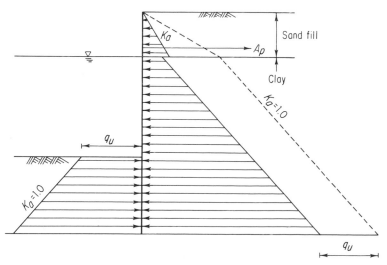

Fig. 9-27 *Pressure diagram for free-earth-support method of design of flexible bulkheads in clay.*

failures are caused by the concentration of loading at the point of anchorage, particularly if the anchorage is relatively rigid. Tie rods have failed because of bending stresses imposed on them by settlements in the soil behind the piling. Because of this, anchor rods are frequently protected against surcharge or fill loads by being encased in boxes. A factor of safety of 3.0 should be maintained against failure of the anchor system. Failure of the soil below the dredge line may be caused by overdredging at some time after completion of the structure. Particularly when a bulkhead has been in place a long time, additional dredging for larger ships may be initiated without realization of the consequences for the existing bulkhead. That is why a conservative approach to the required depth of embedment, such as the equivalent-beam method, is more important than a conservative choice of sheet-pile section.

9-5 SPECIAL RETAINING STRUCTURES

In addition to the relatively simple forms of retaining structures described in this chapter, there are many other structures which retain soil in one way or another, particularly in use in marine construction. Although detailed calculations concerning such special structures fall outside the scope of this book, a brief description will be given of a number of such structures frequently encountered.

Figure 9–28 shows two construction techniques used mainly in wharfs and usually referred to as relieving platforms. The pile-supported platform carries all of the surface loading, which may be quite high on a wharf, and sometimes part of the fill load, thus reducing the pressure on the sheet piling if the latter is placed in front of the platform. If the sheet is placed on the land side of the relieving platform, the main effect of the platform is to reduce the length of sheet piling required. With the sheet piling anchored to the relieving platform, the raker piles under the platform must be designed to support the entire lateral loading on the struc-

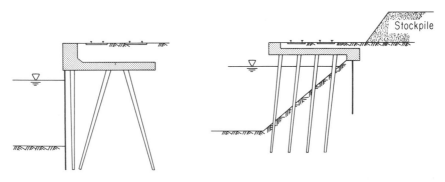

Fig. 9–28   *Two types of wharf design with relieving platform.*

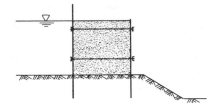

Fig. 9–29 *Double-wall cofferdam.*

ture. The vertical loading of the platform and the fill, transferred to the piles, can be utilized to avoid uplift on some of the piles.

The crib-type retaining wall of Fig. 9–1, which is basically a gravity structure using soil or rock fill for most of its dead weight, has found extensive use in marine work in the form of double-wall cofferdams and cellular cofferdams. A double-wall cofferdam is shown in Fig. 9–29; it consists of two rows of sheet piling connected together with tie rods and filled with soil. It is much used for temporary work in water in the construction of bridge piers, dams, and other foundations. The work area surrounded by the cofferdam can be pumped dry, and the cofferdam must support the water pressure without serious leakage. Since the backfill in the cell preferably consists of relatively permeable soil, watertightness must be provided mainly by the sheet-pile interlocks. Anchors between the sheet-pile rows are short and cheap, and several rows of anchors may be installed, allowing use of light sections of sheet piling. A more sophisticated type of double-wall cofferdam is the cellular cofferdam shown in Fig. 9–30. Here the anchors are eliminated, and the bin pressure of the backfill is retained through tension in the interlocks. An added advantage of the cells is that

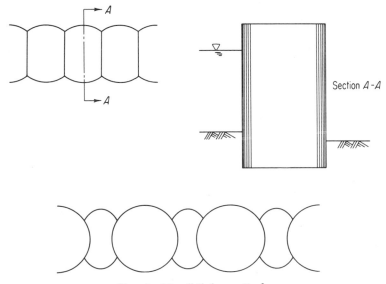

Fig. 9–30 *Cellular cofferdams.*

they are separate, independent units, and therefore no progressive failure need occur after failure of one cell.

Cofferdams must be designed for two systems of loading. In the first place, the assembly must be stable as a gravity-type retaining structure. The total weight of the two walls and backfill must be adequate to prevent overturning due to the water pressure on one side. Although the water level inside the cofferdam can be expected to be inclined from the level on the outside to the level of the excavation, it is advisable to assume buoyant weight of the backfill below the high water level. Resistance against sliding usually presents no problem; an added safety factor is the extra depth of penetration of at least the outside row of sheet piling, which should be adequate to prevent scour damage. To determine the interlock stresses, the method of placing the cell fill and the type of fill should be taken into account. For practically all types of fill with the exception of coarse sand, gravel, or rock, the possibility of at least temporary lack of shearing resistance during deposition in water should be considered. This will cause a lateral pressure equal to the buoyant unit weight of the soil for the fill below the water table when the water table inside and outside the cells is the same, and even above the water table the coefficient of earth pressure may be as much as unity for fine backfills. Therefore, coarse backfill is preferable both for more closely predictable and lower interlock stresses and for improved performance of the cell as a structural unit.

The deflection of such cells under lateral stresses is complicated, and the methods employed in analysis and design are only approximate.

9-6 CONDUITS AND TUNNELS

Conduits and tunnels are structures which must resist both lateral and vertical soil pressures. Conduits are normally laid in excavated trenches and covered with backfill, whereas tunnels are bored in the natural soil or rock by using a variety of means depending on the hardness or resistance of the natural materials. In a sense, this should make the estimation of pressures on a tunnel less complicated than those on conduits, since the tunnel can be considered to take the place of the original soil with a minimum of disturbance. Unfortunately, however, the relation between vertical and lateral stresses in natural deposits is found to vary widely both in soil and in rock, since it depends on the manner of deposition and on the history of loading and tectonic movement since deposition.

The determination of stresses on tunnels in rock is very complicated and far from exact, and it relies mainly on *in situ* measurements using strain-measurement devices. Similarly, the existing stress distribution in soils is estimated on the basis of laboratory triaxial tests and field measurements, possibly with the Ménard pressuremeter (Chapter 7). Strange as it may

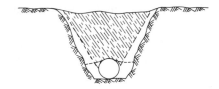

Fig. 9–31   *Pressure on rigid conduit in wide trench with consolidating backfill. Conduit supports cross-hatched area as backfill below top of pipe consolidates.*

seem, the lateral stress at some depth in a natural formation may be higher than the vertical stress both in soil and in rock, since the removal of surface material upon erosion permits vertical expansion and stress decrease, whereas lateral expansion is inhibited. Furthermore, future deformation in a soil deposit (for instance, consolidation under a decrease of the water level) may cause added stresses on the tunnel at a later date.

Conduits in backfilled trenches may be divided into flexible and rigid types. The flexible conduits, when loaded with backfill, compress vertically and become wider, thus increasing the pressure against the backfill on the sides and decreasing the vertical stress. The net effect is generally a fairly even distribution of pressure around the entire conduit. Rigid conduits, however, must support the full weight of the backfill with only very small deflections, and the vertical pressure on the conduit may be in excess of the theoretical overburden pressure of the backfill. The latter case can develop if the trench is considerably wider than the pipe (Fig. 9-31).

Pressure calculation on both rigid and flexible conduits is still largely a matter of empirical rules and local experience. The main precautions which should be taken in all cases are to form a firm, well-compacted bed for the conduit, to compact the backfill well along its sides, and to excavate a trench of minimum width for the conduit's installation.

PROBLEMS

9-1   A mass concrete retaining wall is 15 ft high, 7 ft wide at the base, and 3 ft wide at the top. The rear of the wall makes an angle of 10° with the vertical. The earth retained is granular ($c = 0$, $\phi = 32°$, $\gamma = 120$ pcf^3). The wall is founded 2 ft below the surface of a sandy soil ($\gamma = 125$ pcf, $\phi = 30°$). Backfill in front of the wall footing is compacted to its original density. What are the safety factors with respect to (a) overturning, (b) sliding, and (c) failure of the foundation soil?

Answer   For K_a see Table 9–1. $\beta = 10°$, $\phi = 30°$, $\omega = 0$ gives $K_a = 0.40$. Force on wall is 5,400 lb/ft at 10 ft below the top and makes an angle of $\frac{2}{3}\phi = 20°$ with the normal, or 30° with the horizontal. Weight of wall is 11,260 lb/ft, and resultant force on base (14,600 lb/ft) makes an angle of 19° with the vertical and intersects the base in its middle third. There will be a trapezoidal pressure distribution against the footing. Draw scale model of wall with forces. (a) Overturning: see Fig. 9–5. Take moment about A. Overturning moment from earth pressure: 5,940 ft-lb/ft. Restraining moment from wall weight: 42,600 ft-lb/ft. F.S. = 7.2. (b) Horizontal component of earth pressure: 4,680 lb/ft. If concrete is poured directly on undisturbed soil, friction angle between soil and concrete foundation = ϕ. Restraining forces against sliding:

$$(11,260 + 5,400 \sin 30°) \tan 30° = 8,060 \text{ lb/ft}$$

Also, passive resistance against 2 ft of soil = 750 lb/ft.

$$\text{F.S.} = 8,810/4,680 = 1.89$$

(c) See Fig. 9–22c. $\alpha = 19°$, $\phi = 30°$, $N_{\gamma q} = 5$ for $D/B = 0$, $N_{\gamma q} = 26$ for $D/B = 1$. Interpolation for $D/B = 2/7$ gives approximately $N_{\gamma q} = 12$.

$$\text{F.S.} = \frac{(12 \times 125 \times 7/2)}{(14,600/7)} = 2.5.$$

9-2   What would the same factors of safety be for a vertical, cantilever-type retaining wall with similar height and footing level and a total footing width of 7 ft? The toe extends 2 ft from the face of the wall. There is no key. Both stem and footing thickness are 18 in. Is this design acceptable?

Answer   Earth pressure is horizontal in this case: $K_a = 0.33$, $P_a = 4,500$ lb/ft. Total weight of wall plus soil on toe and heel is 10,400 lb/ft. Resultant intersects base 2.0 ft from the toe, i.e., triangular pressure distribution, and only 6 ft of footing "participates" in bearing. F.S. overturning = 1.94; F.S. sliding = 1.50; F.S. foundation failure ($B = 6$ ft, $\alpha = 23°$, $N_{\gamma q} = 80$) = 1.6. Design, although stable, is not acceptable, since F.S. against failure of the soil below the foundation should be at least 3.0. Footing width must be increased in direction of heel.

9-3   Determine the earth-pressure diagrams for a 30-ft-high cut in a soft plastic clay ($c = 450$ psf, $\gamma = 100$ pcf) using both Peck's lateral earth pressure diagram and Tschebotarioff's neutral earth pressure ratio method. Determine the maximum unit pressure and the total load per lineal foot of cut in each case.

Answer   See Figs. 9–12 and 9–13.

Peck's method: maximum pressure = $\gamma H - 4c = 1,200$ psf.

$$\text{Total load} = \tfrac{1}{2}(H + 0.55H) \times 1,200 = 28 \text{ kips/ft}$$

Tschebotarioff's method: maximum pressure = $0.5\gamma H = 1,500$ psf.

$$\text{Total load} = \tfrac{1}{2}H \times 1,500 = 22.5 \text{ kips/ft}$$

9-4   Make the same comparative calculations as in Prob. 9.3 (a) for a

30-ft braced cut in a medium-stiff clay ($c = 700$ psf, $\gamma = 110$ pcf) and (b) for a 45-ft braced cut in a stiff clay ($c = 1,200$ psf, $\gamma = 120$ pcf).

Answer (a) Peck's method: maximum pressure is 500 psf, total load is 11.6 kips/ft. Tschebotarioff: maximum pressure is 1,400 psf, total load is 21 kips/ft. (b) Peck's method: maximum pressure is 600 psf, total load is 21 kips/ft. Tschebotarioff: maximum pressure is 1,620 psf, total load is 36.5 kips/ft.

9-5 A 31-ft-deep basement excavation must be made through 30 ft of sand and 1 ft into sandstone. Steel-sheet piling is driven through the sand to refusal on the sandstone. As the excavation proceeds, walers are installed at 4 ft and 22 ft below the surface. Knee braces are placed against the walers at 10-ft centers. Excavation and installation of the walers are done in such a manner as not to overload the members beyond their final design load. The upper brace makes an angle of 45° with the horizontal; the lower brace slopes at 22°. The braces are supported by 3-ft square inclined footings, perpendicular to the braces, extending 4.5 ft into the sandstone. The sand properties are $\gamma = 110$ pcf, $c = 0$, $\phi = 30°$. Sandstone properties: $\gamma = 120$ pcf, $c = 4,000$ psf, $\phi = 38°$. The structural design of all members and connections may be assumed to be adequate. Check the stability of the system.

Answer See Fig. 9–11. Trapezoidal distribution gives load of 22.4 kips/ft of wall. The upper brace provides 92 kips of horizontal resistance; the lower brace, 132 kips (graphical analysis). Total axial loads on upper and lower braces are 129 and 143 kips, respectively. Check inclined footings; combine Fig. 9-22b and d. Bearing capacity of upper brace footing: $N'_{cq} = 5.2$, $N_{\gamma q} = 50$, $Q = 268$ kips. Bearing capacity of lower brace footing: $N'_{cq} = 4.5$, $N_{\gamma q} = 25$, $Q = 203$ kips. F.S. for upper brace footing = 2.08, for lower brace footing = 1.42. These safety factors are too low, particularly where cohesion (cementation) and friction of sandstone are combined. If cementation alone were to be used, safety factors would be reduced to 1.44 and 1.13, respectively. There is another reason why the design is not safe. The inclination of the knee braces causes an uplift force of 146 kips on each 10 ft of shoring. The frictional resistance against the sheet piling ($\delta = \frac{2}{3}\phi$) is 82 kips for 10 ft. Since sheet piles were not to be driven deep into sandstone, there would not be any pullout resistance, and the system would collapse.

9-6 The following dimensions are given for a flexible bulkhead in sand: dock surface El. 100.0 ft, water level El. 85.0 ft, dredge level El. 60.0 ft, anchor at El. 90.0 ft, $\gamma = 110$ pcf, $\phi = 30°$. Determine depth of penetration below dredge level, anchor force, and maximum bending moment by (a) free-earth-support method (F.S. = 2.0) and (b) Tschebotarioff's simplified equivalent-beam method.

Answer (a) 15.7 ft, 13,900 lb/ft, 135,300 ft-lb/ft (apply F.S. to K_p). (b) 17.2 ft, 9,400 lb/ft, 95,000 ft-lb/ft.

REFERENCES

1. A. Caquot and J. Kérisel, "Tables for the Calculation of Passive Pressure, Active Pressure and Bearing Capacity of Foundations," Gauthier-Villars, Paris, 1948. English translation.
2. G. P. Tschebotarioff, "Soil Mechanics, Foundations, and Earth Structures," McGraw-Hill Book Company, New York, 1951.
3. R. B. Peck, Earth Pressure Measurements in Open Cuts, Chicago Subway, *Trans. ASCE*, pp. 1008–1036, 1943.
4. G. G. Meyerhof, The Bearing Capacity of Foundations under Eccentric and Inclined Loads, *Proc. Third Intern. Conf. Soil Mech. Found. Engineering*, vol. 1, p. 660, 1953.
5. G. P. Tschebotarioff, "Large Scale Earth Pressure Tests with Model Flexible Bulkheads," Final Report to U.S. Navy, Bureau of Yards and Docks, Princeton University, 1949.
6. P. J. Rowe, A Theoretical and Experimental Analysis of Sheet Pile Walls, *Proc. Inst. Civil Engrs. (London)*, pp. 31–86, January, 1955.
7. H. Blum, "Einspanningsverhaeltnisse bei Bohlwerken," Wilhelm Ernst & Sohn KG, Berlin, 1931.

10 | Soil Drainage and Dewatering

10-1 DEFINITION

The presence of water in soils often poses problems of different kinds for the civil engineer who contemplates construction, excavation, or earth moving. His concern may be with both the free groundwater which can be made to flow through the pores of a soil and the adsorbed and chemically combined water found in largest proportion in fine-grained soils. The latter form of pore water and its effect on the soil properties has been described in Chapter 1, and this section deals only with methods of control of the free water in the soil.

The first problem which comes to mind is excavation below groundwater level. This requires a system of dewatering which not only permits completion of the required excavation but also helps to prevent disturbance and softening of the excavation bottom, softening of the excavation sides, and disturbance of neighboring structures. For the design of a dewatering system it is necessary to know the rate of water flow to be expected. The coarser the soil in the excavation, the higher its permeability and the greater the required capacity of the dewatering plant.

Other problems associated with groundwater are drainage of pavement subgrades to prevent frost damage and drainage of water behind retaining structures, basement walls, and fill embankments to prevent buildup of hydrostatic pressures. A more indirect goal which can be achieved by drainage is the more rapid reduction of pore pressures in fine-grained soil, which speeds up consolidation and causes an increase in strength. The various systems of controlling groundwater and the effect on site and soil conditions will be described in this section.

10-2 RESTRAINT OF FLOW

One of the most direct solutions which presents itself when considering groundwater control is to impose restraints on the groundwater flow. This method is effective in many situations, as is demonstrated in Figs. 10–1 to 10–4. In Fig. 10–1a, flow into the excavation from the permeable upper layer is cut off by closed sheet piling of steel, concrete, or timber. Some seepage from the joints is inevitable, but particularly the rolled steel sections and the tongue-and-groove timber sheet piling can be very effective. The main problem concerned with a solution of this kind is seepage through the excavation bottom with the resulting possibility of developing an unstable floor in the excavation due to seepage gradients or pressures. In clayey soils, where this method can be used without continuous pumping, seepage through the clay "plug" at the bottom is of no consequence, and the remaining question is the thickness of the plug required to avoid instability and heaving of the bottom.

Figure 10–1b shows a method for restraint of flow frequently used in marine work. The impervious earth dike or cofferdam surrounding the site is built up with hydraulically dredged soil, so that flat slopes are necessary because of the loose nature of the material. Only in very rare occasions can a dry condition be maintained without continuous pumping; seepage is most likely to develop through the dredged fill and along the contact line between the fill and the natural soil.

Another method used in underwater construction is shown in Fig. 10–1c. Sheet piling is driven to some distance below the proposed depth of excavation; the excavation is made "in the wet" using clams or other forms of

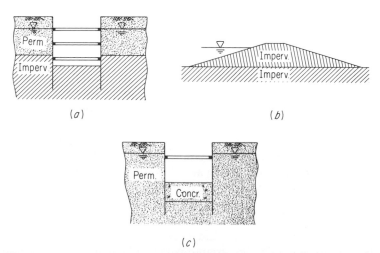

(a)   (b)

(c)

Fig. 10–1   *Restraint of flow into excavation. (a) Sheet piling into impervious subsoil. (b) Dredged cofferdam. (c) Sheet piling and tremie concrete.*

Bentonite slurry   Tremie concrete

Fig. 10–2   *Slurry-trench method of excavation support and groundwater control.*

floating equipment; and a concrete plug or tremie seal is poured below water at the bottom. Once the tremie floor has hardened, it serves as an effective water barrier and the excavation can be pumped out. The system has an added advantage in that the rigid bottom slab acts as bracing for the sheet piling before the full external water pressure is applied. The thickness of tremie concrete required is determined by the head of water outside the excavation; with concrete having a unit weight of about 2.5 times that of water, the minimum thickness of concrete required is roughly two fifths of the head of water against the underside of the tremie slab. Only in relatively narrow excavations can the effect of the bond between the concrete and the sheet piling be taken into account to give a reduction in the required concrete thickness.

A method of restraining flow used mainly to improve the stability of vertical excavations or borehole sides, rather than to create a dry condition, is shown in Fig. 10–2. Instead of being filled with water, the borehole or trench is filled with drilling mud, which consists of a heavy slurry of bentonite or similar material and water. This prevents a gradient developing between groundwater and the hole and reduces a running or caving of noncohesive soils. Although mainly used in boreholes, it can also be used in narrow excavations to support, for example, the walls of sewer, culvert, and subway trenches. Preassembled reinforcement can be lowered in the trench, and concrete can be poured into the mud-filled trench to displace the mud without segregation of the concrete. After the concrete has hardened, it functions as a retaining wall during excavation and construction of the bottom and roof of the tunnel.

Figure 10–3 shows a method for permanent control of water flow under dams by the use of a clay cutoff wall. This method is followed where the dam site is covered with deep layers of alluvium whose complete removal

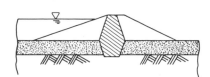   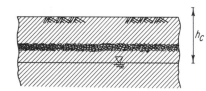

Fig. 10–3   *Compacted clay cutoff wall in alluvium below earth dam.*   Fig. 10–4   *Capillary barrier of gravel or coarse sand interrupting flow of capillary water to surface.*

for the construction of the dam on bedrock would be too costly. In early dam construction, the trench for the cutoff wall was made as narrow as possible and the cutoff wall was made in the wet, using what was called a puddled clay. Subsequent consolidation of this soft material would create a sag in the dam or voids in the cutoff system. This approach is now less common than, for instance, construction of the cutoff wall either as a controlled, compacted fill or as a sheet pile or grout curtain.

In Fig. 10–4 a rather unusual form of groundwater flow restraint, sometimes called a *capillary barrier*, is shown. The fine-grained soil has a capillary head which would cause virtual saturation of the soil between the groundwater table and ground surface. For a number of reasons (frost, construction, agriculture) it may be desirable to prevent this from happening in a surface layer, and a capillary rise can be restrained by placing the thin layer of coarse soil at the desired level. Strange as it may seem at first sight, the same method can be applied to reduce penetration of irrigation water by downward capillary suction in arid regions in order to reduce water requirements or to avoid deep surface-water penetration in heavy soil. Naturally, this would be effective only in cohesive soils, where gravity flow of water is practically nonexistent.

10–3 DRAINAGE THROUGH GRAVITY FLOW

Water control solely by gravity flow is a common feature of highway engineering. Drainage facilities in highways include interceptor drains both for surface-water runoff and groundwater. Only the latter will be described in this book. Most dewatering systems in use in construction projects other than highways, if in permeable soils, are also based on gravity flow of water to one or more collection points, from which it can be pumped to the surface for discharge.

Groundwater control in highway engineering is usually confined to lowering a locally high, or perched, groundwater table under the roadbed to a depth of not more than 5 ft. The lowering is done with a system of subdrains, which consist of trenches containing a pipe drain and filled with pervious material to collect the water. The most primitive form is the so-called French drain, which is a trench filled with gravel or crushed rock. Subdrains of this type were first used not in highway construction but in agriculture to lower the perched water table a few feet in low-lying wet but fertile farm land. To facilitate the discharge, open-jointed clay tile pipe was later placed at the bottom of the gravel-filled trench. Today the tile with open joints is not much in use for important subdrain systems; it has been replaced by pipe of metal, clay, plastic, or fibrous material which is perforated around part of its circumference.

Figure 10–5 shows a typical modern subdrain in impervious soil overlain by a water-bearing stratum. The perforated pipe can be placed

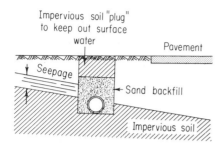

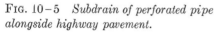

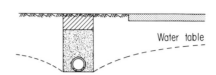

FIG. 10-5 *Subdrain of perforated pipe alongside highway pavement.*

FIG. 10-6 *Subdrain of perforated pipe in relatively permeable soil.*

directly on the impervious soil at the bottom of the trench; placing it on a gravel or sand bed will only cause flow under the pipe, which might give problems at the outlet. A common misconception is that gravel or crushed rock is the best filter material to be used around and above such pipes. Research has shown that gravel filters silt up, or become clogged with fine soil particles, more quickly than a filter consisting of a well-graded sand-gravel mixture.

Figure 10-6 shows a subdrain in permeable soil. Here the drain must be placed at the elevation at which it is intended to maintain the ground-water level. Needless to say, the success of any subdrain system depends on the rate of flow through the water-bearing soil. The section of pipe available for discharge is limited to the part below the lower pipe perforations, and the possibility exists that the quantity of flow that the pipe will carry might be inadequate to substantially affect the groundwater level between drains. In road construction, for example, subdrains parallel to the road are generally laid only under the shoulders. If this arrangement provides inadequate spacing, cross drains are required under the road or the roadbed must be raised to a higher elevation.

Another problem arises when the water-bearing stratum is not permeable enough to yield a significant flow of water into the subdrain, leaving the water table between drains practically unchanged. In this case, unless it is possible to install very deep subdrains on both sides of the road section, the subdrain system will be ineffective. Since water-bearing soils with such a low permeability (very fine sands and silts) are often susceptible to frost heaving and their drainage is highly desirable, other measures such as replacing the soil or raising the roadbed must be considered.

Figure 10-7 presents a different type of drain used in highway cuts, where adverse bedding of rock or soil may combine with groundwater flow along some of the beds to cause a slide. The perforated pipe is installed in holes drilled by augers or diamond drills. Usually such a system of drains, the success of which cannot be definitely predicted before installation, is used only where slides or other signs of instability are observed in existing

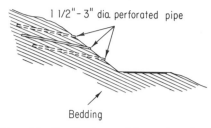

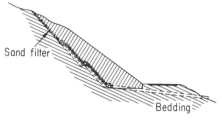

FIG. 10–7   *Drainage of formation above highway cut with drilled-in-place perforated pipe.*

FIG. 10–8   *Support of highway cut with compacted-fill buttress with back drains.*

cuts. A more effective system providing improved stability and water interception is a buttress fill with back drains (Fig. 10–8).

The simplest form of excavation drainage is shown in Fig. 10–9. Water seeping from the sloping sides or entering through the bottom of the excavation is collected in the perimeter trench and pumped out of the excavation for discharge. The seepage from the excavation sides, if significant, will cause erosion and sloughing of the excavation slopes, and the perimeter trench must be cleaned out frequently. Whether a dry bottom can be maintained in this manner depends, as with the highway subdrains, on the spacing between trenches and the depth of the trenches. This system is quite effective when most of the soil in the excavation is relatively impervious and somewhat cohesive, with seepage coming mainly from a few water-bearing layers in the excavation sides. With seepage occurring through the entire section, the maximum seepage gradients and resulting disturbances are at the toe of the slope in the perimeter trench, and the system can be considered for short-time use only.

Proceeding to systems of greater complexity, the next step in excavation protection and drainage is as shown in Fig. 10–10. There the sides of the excavation are protected by closed sheet piling which is braced for lateral support. A sump is excavated at a convenient location near the middle of the excavation (or sumps are spaced at regular intervals), and all bottom seepage is intercepted by the sump. The sides of the sump hole are held up by planking with open joints, gravel banks, or other permeable means of support. To determine if a sump would actually attract all seepage water otherwise escaping through the bottom of the excavation, a rough flow net must be drawn in the manner outlined in Chapter 4. In a borderline case it is always good to remember that in most natural soil deposits the permeability in the horizontal direction is greater than that in the vertical direc-

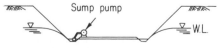

FIG. 10–9   *Excavation drainage with perimeter trench and sump pump.*

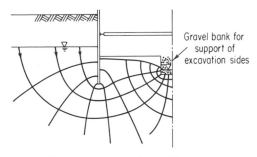

Gravel bank for support of excavation sides

FIG. 10-10 *Excavation with sheet piling and central sump.*

tion, and this will promote flow toward the sump. Another question which must be resolved is the possibility of *piping* in a sump bottom. It will be recalled from Part I that a piping, boiling, or quicksand condition develops when the upward seepage force in the soil equals the weight of all material above that point. The equation for the critical gradient in which this condition develops was given in Chapter 4:

$$i_c = \frac{G_s - 1}{1 + e} \tag{4-27}$$

The flow net will give the hydraulic gradient below the sump bottom, and the possibility of piping can be checked. If piping appears to be a possibility, the sump can be "weighted down" with what is called an inverted filter (Fig. 10-11). The filter consists of layers of successively coarser material from the bottom of the sump pit up; its purpose is to permit the passage of water but to prevent soil grains from moving upward. In this case the sides of the sump pit are not utilized for water infiltration and all seepage is directed toward the bottom. The weight of the filter material will prevent boiling of the natural soil in the sump bottom, but to be effective the filter must meet two requirements in addition to a certain weight. In the first place, it must be coarse enough that the loss of head during flow through the filter, and hence the seepage forces in the filter, will be relatively small. If this is not the case, the

FIG. 10-11 *Inverted filter to avoid piping in sump.*

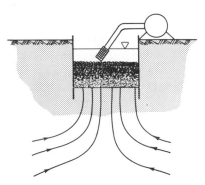

piping problem is simply moved upward a couple of feet from the natural soil to the filter material. In the second place, each layer of the filter must be fine enough to prevent passage of fine-grained material from the underlying layer through its pores.

The first condition can be readily met and an ample margin of safety against piping can be obtained by using soil in the size range of coarse sand or gravel. The second condition obviously is dictated by the grain size of the natural soil in the sump bottom. Were the soil and the filter material built up of closely packed spheres of uniform size, the smaller spheres would be able to pass between the larger spheres when the latter's diameter was more than 6.5 times that of the smaller spheres. Because of the angularity and nonuniformity of soils this ratio of sizes can be higher in practice before fines begin to wash through. A well-tried rule, first proposed by Terzaghi, for filter material is

$$\frac{D_{15} \text{ of filter}}{D_{85} \text{ of foundation}} < 4 \text{ to } 5 < \frac{D_{15} \text{ of filter}}{D_{15} \text{ of foundation}} \tag{10-1}$$

The first inequality of the above equation is designed to prevent fines of the foundation material from washing through the filter: the 15 percent size of the filter material (or the grain size which is greater than 15 percent by weight of all filter particles) must be less than 4 to 5 times the 85 percent size of the foundation material. The second inequality has the function of keeping the seepage force comfortably low. For continuity of flow, the velocity of flow will be the same for the sump foundation material as for the filter. According to Darcy's law, therefore, the gradient will be inversely proportional to the permeability. Experiments have shown that the permeability of soils is roughly proportional to $(D_{15})^2$, and consequently the seepage gradients and forces in the filter material will, by the above requirements, be one-sixteenth to one-twenty-fifth of those in the sump bottom material. The filter can be built up in layers, each layer being coarser than the one below in accordance with the above equation.

The use of filters is discussed here very briefly for application in dewatering problems in construction, where filter design and installation form a relatively unimportant phase of the overall work and where a hit-or-miss method of installation is frequently followed when difficulties develop in the excavation. Filters play a much more important role in earth-fill dams, where no high flow gradients can be permitted at the downstream end if stability of the toe is to be assured.

The next step to more complicated dewatering systems based on gravity flow is the installation of wellpoints (Fig. 10–12). An individual wellpoint is a 2- to 3-in.-diam, 12- to 42-in.-long perforated pipe provided with a brass or stainless-steel well screen and connected to a vertical pipe or *riser*. The points are placed in a row or ring, and the riser pipes are attached through a common manifold or header pipe to a special wellpoint pump.

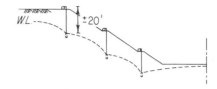

FIG. 10–12 *Three-stage wellpoint system.*

With this suction pump it is possible to bring water to the surface from a maximum depth of about 20 ft. The spacing between the wellpoints is usually 3 to 5 ft, but it can be as much as 10 ft, depending on the reduction of groundwater level required. For dewatering excavations which are more than 20 ft below the water table, a multiple-stage system of wellpoints can be used, as shown in Fig. 10–13. The excavation can proceed to just above the level of the previously installed row of wellpoints, and the next row can be installed in the dry excavated area and pumped before excavation proceeds.

A wellpoint system is installed by first laying out the 6- to 12-in. header pipe. The wellpoints on the riser pipes are then inserted either by jetting or by placing them in drilled holes. The jetting process is similar to that described in Chapter 7 for advancing a borehole. The self-jetting well-

FIG. 10–13 *Two-stage wellpoint system next to creek.* (Courtesy American Dewatering Corporation)

point tips available with all commercial wellpoint installations are very efficient. In soils containing fine-grained materials, another filter is desirable, and this can be provided by adding concrete sand as the point is jetted down, so that the wellpoint tip is finally surrounded by an annulus of sand. The sand also helps in directing drainage to the wellpoint. The hole made can be enlarged by a "jetting chain" running up the pipe or with a special, large-diameter, star-shaped tip. No filters are needed when dewatering clean sands.

Operation of a wellpoint system requires round-the-clock attendance; for an interruption of pumping can have catastrophic consequences, particularly when the system is used not only to lower the water table in an excavation but also to reduce pressure on the shoring as described later in this section. One auxiliary pump should always be available for each two pumps in use, and one for each pump in critical cases.

Another method used to lower the water table to depths greater than 20 ft by gravity flow is the deep-well system (Fig. 10–14). The 6- to 24-in. casing is provided with a long screen (20 to 75 ft), and in this case a submersible pump with the ability to push the water up to a height of 100 ft or more is installed near the bottom of the well. These deep wells can be installed in rows, and because of their greater reach in depth and capacity of discharge the spacing can be much greater than that for wellpoints, ranging from 20 to 200 ft. Although each well has its own pump, in most cases the system is quite efficient; for the submersible electric pumps available today for the smaller-size wells are very reliable and require little attention. Occasionally this advantage is offset by union requirements that a number of electricians attend the pumps. The large, high-capacity wells usually have pumps with the motor at the surface.

Deep wells such as these are installed with rotary drilling rigs or other equipment available for making large, cased holes. No drilling mud can be used, because it will tend to seal off the borehole sides. Preferably only

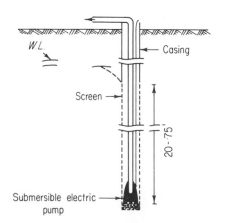

FIG. 10–14   *Deep well with submersible pump.*

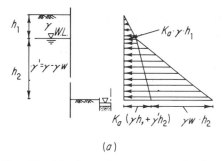
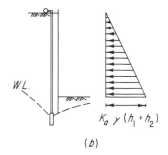

(a) (b)

Fig. 10–15   *Lateral pressure on shoring of excavation with (a) dewatering inside shoring and (b) dewatering by wellpoints outside shoring.*

water is used with sufficient head to prevent squeezing or caving of the hole. The screen must be centered carefully in the hole to assure an equal amount of filter material around the screen perimeter. The holes must be made vertical and straight, since pumps and riser pipes must be lowered down them. The consequences of a temporary interruption in the pumping of one deep well in a row is less serious than that of an entire wellpoint stage in most cases, and less auxiliary equipment is needed.

The great depths of dewatering made possible with the wellpoint and deep-well systems open possibilities for dry excavation and construction which did not exist with conventional sump drainage. The latter system could begin to operate only after the excavation was completed, whereas the deep systems permit lowering of the groundwater table before excavation. This is a tremendous advantage in all noncohesive soils, where even small seepage gradients will cause some caving and sloughing of the excavation sides and softening of the bottom. Deep wells or wellpoints can in some cases eliminate the need for shoring (Fig. 10–13) and can reduce the pressure on shoring where shoring is required by space limitations. In addition to reducing pressure on the shoring by 50 percent and more as shown in Fig. 10–15, the type of support required no longer has to be watertight and the depth of its penetration below the excavation can be reduced also.

Another case where deep dewatering is essential is indicated in Fig. 10–16. The artesian pressure head in the sand below the impervious clay layer would be high enough to push up or break through the reduced

Fig. 10–16   *Deep-well drainage prevents possible failure of excavation bottom as the result of artesian pressure head.*

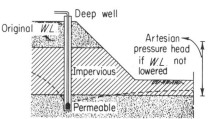

height of clay in the excavation. Even though no sign of groundwater may have been observed at foundation level, it is necessary to reduce the head below the excavation with a deep-well system. This example emphasizes the importance of an adequate depth of preliminary investigation.

10-4 DEWATERING WITH FORCED FLOW

Very fine sands and silts and sand-gravel mixtures with a substantial silt content are unable to release a significant amount of water through gravity flow only. The capillary action which may result in a capillary rise of 2 to 5 ft in silty sands and up to 20 ft in fine silt retains the water, and the flow into any of the previously described dewatering systems would be negligible. However, such soils usually have little or no cohesion or cementation, and their stability both in the sides and in the bottom of excavations below the water table can be seriously affected by saturation. Moreover, loose saturated fine soils of this type are very sensitive to sudden changes in stress; the shearing stresses caused by the vibrations or impact loading of construction equipment may develop an increased pore pressure leading to subsequent liquefaction of the mass.

For successful dewatering in these fine, noncohesive soils it is necessary to apply a suction head in excess of the capillary head to the dewatering system. The pumping capacity of the system does not have to be great, because of the low permeability of the soils. The permeability of a silty sand may be as low as 10^{-5} cm/sec compared with a permeability coefficient of 10^{-1} cm/sec in a medium sand. This implies that the dewatering points must be at relatively close spacing, and the distance between the sides of an excavation and the dewatering points should be kept as short as possible.

Both the wellpoint system and the deep-well system can be adapted for dewatering fine, noncohesive soils by maintaining a vacuum in the well with the use of airtight seals for all joints (Fig. 10-17). In a conventional wellpoint system the vacuum head that can be obtained is relatively small, since the maximum suction head at the header line is only 20 ft. Much better results can be obtained with deep wells fitted with long screens and submersible pumps and a vacuum line connected to the top of the well.

For even finer soils, in the fine silt and clay size ranges, an entirely different method of dewatering, an electrical method, has been developed by L. Casagrande.[1] If an electric potential field is applied to a saturated soil—for instance, by inserting two electrodes some distance apart in the soil—the cations in the pore liquid and hence the pore liquid itself will move toward the negative electrode or cathode. If a well or wellpoint is connected to the cathode, this water can be withdrawn and the flow will continue. The anode may be any nearby available conductor such as a

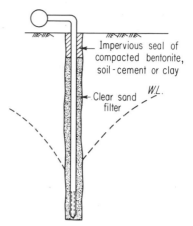

FIG. 10-17 *Wellpoint with filter and seal for vacuum dewatering.*

pipeline or railroad track. This process is called electroosmosis. The velocity of flow v_e toward the cathode can be expressed as

$$v_e = k_e - i_e \tag{10-2}$$

where k_e is an electroosmotic coefficient of permeability and the term i_e, equal to E/L, the electric potential divided by the distance between the electrodes, is an electric gradient. The electroosmotic permeability differs from the previously defined hydraulic permeability in that it is independent of the size of the pores but is proportional to the porosity n of the soil, besides varying with the electrolytic and viscous properties of the fluid. The porosity of almost all soils varies within a relatively narrow range, and it is therefore found that the electroosmotic coefficient of permeability is roughly independent of soil type, since the electrolytic properties of pore water are also relatively constant. For an electric potential gradient of 1 volt/cm, the electroosmotic coefficient of permeability can be taken as 0.5×10^{-4} cm/sec for most soils. For this reason, with practical electric and hydraulic potentials, there is a size range of soil (silt) above which hydraulic drainage is more effective and below which electrical drainage becomes preferable.

Since electroosmotic dewatering is quite costly and its use is confined to finer-grained soils where the actual quantity of drainage presents no difficulties, it is employed mainly in situations where the secondary effects of dewatering, such as consolidation and the increase of shear strength, are of importance. The technique has been used to increase the bearing capacity of piles in fine soils by using the pile as the cathode.[2]

Dewatering for the purpose of consolidation and shear-strength enhancement in soft, fine-grained soils can also be achieved by the different technique of preloading the soil. To accelerate the pore-water flow and the rate of pore pressure dissipation under a preloading surcharge, the sand drain system has been developed. Vertical sand drains consist of a

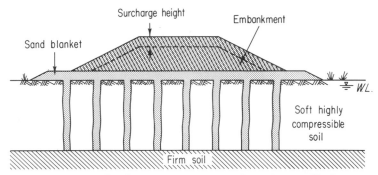

Fig. 10–18   *Sand-drain system for accelerated consolidation.*

series of holes drilled or punched in the fine-grained soil and filled with free-draining, coarse granular soil. The vertical holes or drains are installed in a regular pattern, and they are usually connected at the surface by a drainage blanket of permeable soil which forms part or all of the surcharge (Fig. 10–18). The excess pore pressure which causes flow to the drains is provided by the surcharge load. The flow, which normally would be directed to the top and bottom of the formation, is accelerated tremendously, since the maximum distance of flow for the pore water is reduced to one-half the spacing between sand drains.

Sand drains are used mainly in thick, highly compressible deposits of organic soil, silt, or clay which must function as foundation material for road and embankment fills. In the placement of high embankment fills, the use of sand drains can result in a higher rate of construction without shear failures. To prevent instability, the rate of loading must be carefully controlled with frequent reference to pore pressure and settlement readings. In the case of fills of low height, interest is usually centered in settlement rather than possible failures and the rate of settlement can be speeded up by placing, temporarily, a height of fill greater than with which the site will ultimately be loaded.

Sand drains usually have a diameter of 18 to 24 in. and are placed at 6- to 10-ft centers. Installation is carried out in most cases by driving a closed mandrel into the soil with a pile driver. The bottom plate can be pulled up or left in the hole, and sand is added as the casing is pulled up. The installation is by no means cheap, and the advantages must be weighed carefully against the cost.

10–5 INVESTIGATION

When a routine soil investigation reveals the possible need for a dewatering system, additional field work is required to determine the most suitable system of dewatering and to establish design criteria for the system. The most important information which should be obtained is:

1. Elevation of groundwater and possible fluctuations
2. Proximity of sources of free water
3. Detailed soil profile, including classification and grain-size distribution of soils found below the groundwater table
4. In-place permeability of soils below water table
5. Effect of dewatering on proposed structure and, if applicable, adjoining structures

The necessity of accurate groundwater observations was stressed in Chapter 7. It should always be remembered that the groundwater table is subject to seasonal and other fluctuations. Well records or information from other excavation projects in the area may be useful in the evaluation of the extent of such fluctuations. If the site is near a free body of water, variations in the groundwater level are likely to be small. The proximity of a free body of water will usually have a considerable influence on the capacity of pumping plant required.

The chart of Fig. 10–19, which shows the possible cases and limitations of various types of dewatering equipment, clearly demonstrates the importance of an accurate profile of soil conditions to a depth well below the proposed depth of excavation. It is particularly important to know if certain strata of soil are continuous over the full area of the site. The presence of a continuous clay or silt seam a few inches in thickness just above the proposed level of a row of wellpoints in a sandy soil has a very great influence on the efficacy of the system. Such a seam must be "bridged" with an adequate sand filter, which cannot be placed economically after the system is found to be ineffective without it. Similarly, a

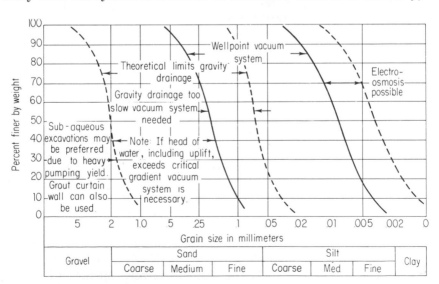

Fig. 10–19 *Grain-size limits for soil-dewatering methods.* (Courtesy Moretrench Corporation, Rockaway, New Jersey)

sump system may fail to intercept all water below the excavation bottom if the soil profile contains lenses of impervious soil such as often occur in alluvial fans and glacial deposits. Most dewatering operations are very costly, and a considerable part of the cost is in the initial installation. It is the soil engineer's responsibility to make sure he has enough information to predict the successful performance of the system selected. A chart as shown in Fig. 10–19 may be useful for a preliminary selection of a dewatering system.

In many cases the soil and groundwater conditions as observed during the field investigation do not provide a clear-cut answer as to the most suitable method of dewatering required. This is particularly the case in fine soils where the depth of excavation will be less than 5 to 10 ft below the water table. In the boreholes, seepage may be insignificant because the hydraulic head is low and the borehole sides are smeared during drilling, forming a thin annulus of low permeability around the hole. However, conditions in the full-scale excavation may be such that sump drainage cannot cope with the flow and a wellpoint system may be required. In particular, if relatively high allowable foundation pressures are recommended for the soil at the bottom of the excavation, no disturbance from upward water seepage can be risked. If the proposed excavation depth is not great, say 10 to 15 ft, it will frequently be worthwhile to verify excavation conditions with a few exploratory trenches dug with a backhoe.

In his selection and design of more complex dewatering systems, the engineer cannot safely rely on the permeability coefficients of the soil as determined in the laboratory from tests on more or less undisturbed samples. The laboratory test results vary too much from the *in situ* values, particularly in stratified or otherwise oriented deposits. Consequently, the permeability of the soil should be estimated in the field by means of pumping or seepage tests. The ideal pumping test would be a small-scale operation of the type of plant (wellpoint, deep well, sump) anticipated for use. Except for large projects this is not usually practicable, and pumping tests or seepage tests must generally be carried out in the existing boreholes instead. Several common methods are described in the examples at the end of this chapter.

After adequate preliminary testing leading to the design and installation of the dewatering system, successful system operation can be determined in many cases quite simply, of course, from the state of the excavation. In some of the examples given previously, however, it would not be wise to proceed with the excavation without first checking to see if the system performs in accordance with the expectations. For instance, in the case of Fig. 10–15, where the wellpoint system is relied upon to reduce the shoring pressure, the lowering of the hydrostatic plane behind the shoring must be checked after the system has been operating for some time and before excavation has begun. This checking can be done quite easily by installing dummy wellpoints, i.e., wellpoints not connected to the pump, to act as piezometers in following the decline of the ground-water table for some distance away from the system. They should also be provided with a filter to prevent clogging by fine-grained soils.

In very fine soil, where forced-flow dewatering is used, a piezometer in the form of a wellpoint and riser pipe could not collect enough water to reflect the hydrostatic pressure or pore-water pressure in the soil. For this purpose special piezometers have been developed; they consist of fine, porous material, and they require very little flow of water to give a reading. The most familiar of this kind is the Casagrande porous-tube piezometer (Fig. 10–20). A more advanced type, developed by the U.S. Bureau of Reclamation, is shown in Fig. 10–21. The double tubing on the latter model serves to flush out air bubbles, which are the main source of error. Many other pore pressure-measuring devices have been developed both for use in inside borings and as probes which can be inserted without drilling. If the head in the pore water is below ground surface, it can be gaged by lowering a small electric probe on a wire into the riser pipe, which has an inside diameter of $\frac{1}{2}$ in. or less. By flashing a light or operating a buzzer, the device indicates when the water surface in the piezometer is reached. If the pressure head is greater than the elevation of the observation point, bourdon or other gages can be used.

An electrical pore-pressure-measuring device has been developed in the Netherlands.[3] The flow of pore water needed to record pressure changes is limited to that inside the narrow entrance channels to the piezometer, and the pressure is recorded electrically through a membrane with a strain gage. Very rapidly changing pressures, such as from traffic on highways or runways, can be measured with this device and automatically recorded.

The remaining question to be investigated is the effect of the dewatering

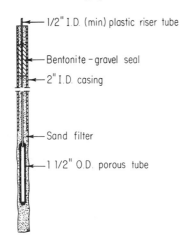

FIG. 10–20   *Casagrande   porous-tube piezometer.*

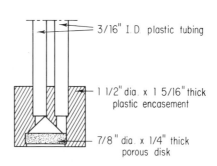

FIG. 10–21   *U.S. Bureau of Reclamation piezometer with double tubing for flushing out air bubbles.*

on the proposed construction and possibly on adjoining existing structures. Dewatering increases the effective unit weight of the soil originally below the water table by 62 pcf, which can cause substantial settlement both in the dewatered zone and in the soil below. Inside the excavation this effect is generally more than compensated for by the removal of soil, but just outside the excavation the net increase in pressure cannot be ignored if other structures are within the area of influence of the system. If a significant amount of settlement is to be expected, modifications to the dewatering system must be considered. Excavations with closed sheeting and sump drainage do not affect the groundwater table outside the excavation to any noticeable degree. The influence of wellpoint or deep-well installations on the surrounding groundwater conditions can be practically eliminated with watertight shoring on the outside of the system. Another solution sometimes used is to pump part of the discharge into a system of recharge wells adjoining the existing building. This has been done quite often in old cities in Europe, where many of the existing buildings are founded on timber piles which must not be left exposed to air.

Inside the excavation, the question to be resolved concerns the stage of construction at which dewatering may be stopped. The excavations for footings must be backfilled before water can be allowed to seep up around the footings. If a basement has been constructed, the weight on the foundations must be more than adequate to prevent floating when dewatering is stopped. If dewatering is also used for pressure release on shoring, backfilling must be completed to a height somewhat above the original water table. This is good practice in most cases, since backfilling under water cannot be controlled properly, and settlement of roads and walkways may develop after completion of the contract.

10-6 SECONDARY EFFECTS OF DEWATERING

Although the seconadry effects of dewatering have been mentioned briefly in the preceding pages, they are of enough importance to warrant a separate discussion. These side effects, whose possibly harmful consequences must always be checked when dewatering is required, may be so beneficial in other instances that they justify the cost of a dewatering system although no excavation below groundwater is contemplated. This will be demonstrated with a few examples.

Preconsolidation. When construction on a formation of high compressibility is contemplated, it is usually decided that a pile foundation or other deep foundation is the best solution to avoid settlements. In some cases, however, this is not economically feasible and instead the area is surcharged with a load whose pressure is equal to or greater than the design pressure. This causes settlements before construction which otherwise would occur under the actual structure. There will usually be some elastic rebound when the surcharge is removed and some recompression when the design load is applied, but this is rarely more than a small fraction of the initial settlement. Surcharging is normally done with earth fill, and sometimes this itself is an expensive operation. If the soil near the surface lends itself to wellpoint dewatering (silt and sand grain-size range) and if the water table is close to finished grade, it may be more economical to place a ring of wellpoints around the area to be surcharged and to lower the water table by 10 to 15 ft. This has the same effect as a 6- to 9-ft layer of fill on the soil below the wellpoints, and the seepage and capillary forces will cause appreciable densification in the upper layer. If the water level is at existing ground level, the system has the added advantage of temporary hardening and improving working and traffic conditions at the site. The sand drain system already mentioned is a different case, where the required period of surcharging is reduced substantially by improving subsoil drainage.

Shear-strength Gain. The system of surcharging has long been popular for improving shear strength of cohesive soils. The gain in shear strength is a direct consequence of forced-flow dewatering, or dissipation of pore pressure, causing an increase in effective vertical stress and a corresponding increase in shear strength along a possible plane of sliding. The sand drain method has made this method attractive in cases where without drains it would be too time-consuming. In sensitive clays, however, it is possible that the shear strength will decrease initially upon installation of the sand drains, particularly when a closed-end mandrel is used.

PROBLEMS

10-1 Figure 10–22 presents a pumping trial for a dewatering project. A test well is driven to the top of the clay layer; and after the pumping

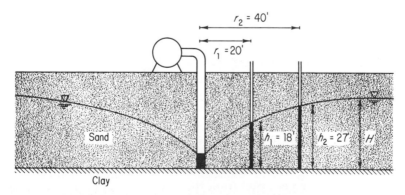

FIG. 10–22 *Pump test for permeability determination.*

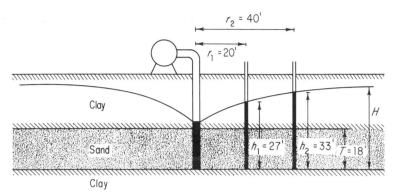

FIG. 10-23   *Pump test in confined sand layer.*

reaches a constant discharge Q, the water level is measured in the two observation wells. If $Q = 0.5$ ft³/sec, what is the coefficient of permeability k of the sand?

Answer   Discharge can be expressed as $Q = kiA$. $i = dh/dr$, and $A = 2\pi rh$, so $Q = k\,2\pi rh\,dh/dr$. Integration between h_1 and h_2 gives $k = [Q/\pi(h_2{}^2 - h_1{}^2)]\ln(r_2/r_1)$. This gives $k = 2.7 \times 10^{-4}$ ft/sec.

10-2   What is the value of k in the case of Fig. 10–23 if Q is 1 ft³/sec?

Answer   In this case, $A = 2\pi rT$, $i = dh/dr$, $Q = k\,2\pi rT\,dh/dr$. Integration gives $k = [Q/2\pi T(h_2 - h_1)]\ln(r_2/r_1) = 1.02 \times 10^{-3}$ ft/sec.

10-3   For the excavation shown in Fig. 10–16 the original water level was 25 ft above the excavation bottom. The stiff clay ($c = 2,000$ psf, $\gamma = 125$ pcf) layer at the excavation bottom is 8 ft. A blackout of long duration stops the pumps and causes serious concern for the excavation bottom. The superintendent correctly decides to fill the excavation with water to avoid collapse of the clay layer and subsequent inundation of the excavation with sand-bearing water. How high should the water reach inside the excavation to maintain an equilibrium?

Answer   The clay strength has no significant influence. The clay blanket would fail in bending, and the strength of the clay must be neglected. The excavation must be filled with at least 16 ft of water.

10-4   A basement excavation is made in sand, 15 ft below the groundwater surface. The excavation is kept dry with wellpoints. Several large column footings are completed when a pump failure causes flooding of the excavation, and some sand flows in from sides and bottom. After the water is lowered again, the excavation is cleaned up and the foundations are completed. As the building is being completed, the foundations which were in place prior to the pump failure show excessive settlement. Explain.

Answer   As water flowed into the excavation through its sides and bottom, a high hydraulic gradient developed under the footing edges. (Try to sketch a flow net.) Soil from the finer fraction below the footing perimeter was carried off, but subsequent deposition of mud camouflaged the damage.

REFERENCES

1. L. CASAGRANDE, Electro-Osmotic Stabilization of Soils, *J. Boston Soc. Civil Eng.*, January, 1952.
2. L. G. SODERMAN and V. MILLIGAN, Capacity of Friction Piles in Varved Clay Increased by Electro-Osmosis, *Proc. Fifth Intern. Conf. Soil Mech. Found. Engineering*, vol. 2, 1961.
3. G. A. PLANTEMA, Electrical Pore Water Pressure Cells: Some Designs and Experiences, *Proc. Third Intern. Conf. Soil Mech. Found. Engineering*, vol. 1, p. 279, 1953.

11 | Soil Property Modification and Pavement Design

11-1 INTRODUCTION

The human mind is constantly searching for new ways to improve on existing conditions, and soil conditions are no exception. Since people began traveling, they have been confronted with the frequently undesirable conditions in which nature has left the terrain; and as wheeled traffic developed and began to increase, efforts were made to build roads and to improve the existing soil conditions. With the known troubles associated with soft ground, the early roads consisted of a hard surface placed over the natural ground. With time, the technique of compacting or densifying layers of soil before placing the pavement developed. The compaction is achieved by running heavy vehicles with special wheels or rollers over the soil. This is sometimes termed mechanical compaction or stabilization.

At first, compaction was limited to a thin surface layer, but cutting and filling to create acceptable road gradients soon involved compaction of fills of considerable thickness. From this point the development of a specialized trade occupied solely with excavation and compaction of soils has been very rapid, especially in the United States. It is of interest that, notwithstanding the tremendously increased costs of labor and equipment and the vastly improved and upgraded standards for compaction, the cost (reduced to a standard reference) per cubic yard of excavating and recompacting soil has declined steadily since the beginning of the century. This fact, in combination with the high cost of building and the increased value of land, has led to the construction of large earth embankments where trestles would have been used before, to complete changes in topography by cutting off hilltops and filling in valleys to create level land for development, and to construction of earth-fill dams of enormous dimensions.

Compaction equipment has been improved and increased in size both to handle the large quantities of fill material at the same rate as modern grading equipment can move it and to attain standards of compaction which have been raised to accommodate the increased frequency and weight of vehicular traffic or to support the great loads of modern aircraft. The depths of fill in high dams and in developments in hilly terrain also demand a higher degree of compaction to minimize settlement of the fill under its own weight. Construction on land developed by extensive grading of steep ridges and deep valleys may result in a structure founded in recently cut bedrock on one side and on over fifty feet or more of compacted fill on the other side. Naturally, the bearing properties of compacted fill and its foundation should be as nearly as possible similar to those of the bedrock in the cut.

In addition to mechanical compaction of soil, methods have been developed to stabilize soil by blending it with different soil or other materials. For instance, the grain-size distribution of a soil in place may be so poor that no amount of compactive effort can give the soil the necessary strength and stability. Blending with other soil can solve this problem quite economically in many cases. This is done frequently in earth-fill dams and also in highway construction where susceptibility to frost damage, the drainage properties, and the strength after compaction can be modified by mixing different soil types. In some cases, the quality of a natural deposit for compacted fill can be improved by removing certain grain sizes, as by washing gravels used in pavement base and by removing boulders from soil fills. The strength of compacted fills can be increased chemically to a considerable degree with small quantities of cement, asphalt, lime, or other additive. This course is followed particularly in highway and airfield pavement construction, and a description of this form of soil stabilization is a necessary preamble to the second part of the chapter, which covers pavement design. It will be recognized from the discussion so far that the construction material, concrete, represents a chemically stabilized soil whose grain-size distribution has initially been carefully controlled. The grain-size distribution and amounts of cement and water used in concrete are all based on the results of research to find the effects of variation in these quantities on the concrete's strength.

11-2 SOIL COMPACTION

The compaction of soil to a higher density is an important phase in the construction of airfields and highways. It is also the only way in which large embankments, earth dams, and land-development fills can be given enough strength and density for satisfactory lifetime performance. Compaction also offers the unique opportunity of preparing a soil material to suit site conditions, a soil far more consistent in its engineering properties than most natural deposits. This permits the soil engineer to be less conservative in his estimation of the field behavior of such a compacted fill on the basis of laboratory experiments than he could be with most natural deposits. However, as with most man-made products, the quality is directly related to the workmanship of the manufacturer and to the intensity of control exercised. As can be imagined, lack of adequate control over compaction and a resulting inadequacy or nonuniformity of the engineering properties of the material can be catastrophic in the case of dams and other high fills.

One of the first discoveries in connection with soil compaction was the relation between the moisture content of a given soil and the degree of compaction or the unit weight which can be obtained with any one form of compaction equipment. Figure 11–1 shows the relation between moisture content and dry unit weight of different soils compacted in the same manner. The method of compaction happens to be a laboratory one, but similar graphs can be obtained from field compaction results. It is evident from the shape of each curve that each soil type has a moisture content at which the maximum dry unit weight can be obtained with the particular method of compaction used. This is called the optimum moisture content. The shape of the different curves indicates that some soils are more "moisture sensitive" than others; the effect on the dry unit weight of a change of moisture content is much smaller, for instance, in the case of a

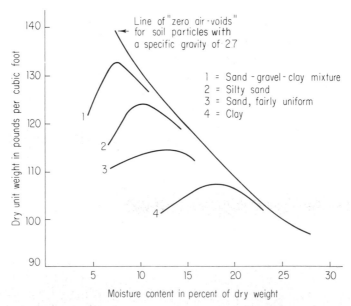

FIG. 11–1 *Moisture content–dry unit weight relationship for different soil types compacted in the same manner.*

sand than a sand-gravel-clay mixture. This means that a much more rigid control is needed over the moisture conditions in clay fills than in sand fills.

The curve of dry unit weight versus moisture content for a given soil may be quite different when other field or laboratory methods of compaction are used. The optimum moisture content, however, varies within relatively narrow limits for different compaction methods for most soils. The exception is sand, which can be vibrated to a high density while dry, whereas when rollers and tampers are used, the sand must hold a reasonable amount of moisture before it will again attain a maximum unit weight under compaction.

The moisture content having been established as the most important influence on compaction results, the remaining factors of influence will be discussed: the type of equipment used, the thickness of successive fill layers compacted, and the number of applications of compactive effort on each layer.

Compaction Equipment.   The compaction equipment in use today can be divided into three different mechanical movements: tamping, rolling, and vibrating. Tamping is used mainly for compacting relatively small quantities of fill which is not accessible to larger compaction equipment, such as backfill behind walls or around footings and fill under floor slabs inside buildings. The tampers most commonly used are the "frog"-type tamper which is operated by one man who makes it jump up and down by firing a gasoline cylinder. The weight varies from one ton to less than a hundred pounds. Large, mobile equipment featuring a row of tampers on a vehicle is in use for compaction of soil-cement and gravel beds under railways. The tamper can be used with success in most soils which do not contain large stones, but in silts and fine sands with relatively high moisture contents there is danger of local liquefaction. This phenomenon is similar to that which can be experienced by jumping up and down on fine beach sand just above the waterline. Tamping of fill is relatively inefficient; the layer of soil which can be compacted is usually 4 in. or less, and the surface area of the tamper is rarely more than 3 ft^2.

Compaction of fill by rolling is by far the most common method in all fills which are accessible to mobile equipment. This has been the case since the days of the steamroller, which for many years was a familiar sight on every highway project and which is still in use in many countries. In the United States, the smooth-wheel rollers, now driven by gasoline engines, today are used mainly for compaction of crushed stone, crushed slag, or other base course materials and for compacting asphalt pavements.

In soil compaction, the sheepsfoot roller (Fig. 11–2) and similar steel-wheel rollers with footprint pressures ranging from 50 to 500 psi have replaced the smooth-wheel rollers. Light sheepsfoot rollers range in weight from 5,000 to 10,000 lb, and heavy rollers up to 75,000 lb are being used.

Fig. 11-2   *Sheepsfoot roller pulled by dozer. Spraying is required to maintain optimum moisture content.*

Most rollers are towed by other equipment, but self-propelled sheepsfoot rollers are also available. In addition to the sheepsfoot, other patterned wheels have been developed, featuring both larger and smaller feet. Each of these wheel types has been developed for particular conditions which may be prevalent in a local area. For instance, a roller featuring spiked feet with high footprint pressures can be very useful for breaking up chunks of shale and other soft rocks in an earth fill, while rollers with relatively large feet are used in soils which are easily kneaded. The two most important properties of the equipment are the weight and the footprint pressure; they determine the thickness of fill which can be compacted in one single layer and the number of passes which have to be made for each layer for some specified amount of compaction of the soil involved.

Rubber-tired rollers have been in use for over thirty years. The first of this kind, still in use on many highway projects for compacting thin layers of granular base course materials, is the wobble-wheel roller. It has two rows of rubber-tired wheels which are mounted at a slight angle with the axles, and it has a container which can be loaded with ballast to give it a gross weight of 8 to 12 tons. Rubber-tired rollers up to 200 tons in weight with tire pressures up to 150 psi have been developed since, mainly for compacting subgrades and base courses for airfield pavements. The increased weight of aircraft and the high-pressure tires used make very severe demands on the compaction of the soil in such circumstances. On other compaction work the large rubber-tired rollers are also finding more use, but the steel-wheel rollers are still the most common sight on large earth fills.

Another method of compacting soils is by vibration. Originally this technique was developed for compaction of relatively dry, sandy soils.

The success of the vibrating probes and needles in concrete work led to use of the same equipment for compaction of hard-to-get-at backfill behind retaining walls. This amounted to densifying noncohesive soil by shaking it long enough to allow the particles to arrange themselves in a dense matrix with the aid of gravity. The next step was to apply pressure to the fill while vibrating with surface vibrators. In this application hand-operated vibrating plates have come to be used extensively for compaction of granular fills inside buildings. For larger projects vibrating rollers which used both smooth and patterned steel wheels were developed. These rollers, which have a vibrator mounted over the wheel, have been very successful not only in dry, noncohesive soil but in many cohesive fills also. They have been very effective in fills containing large amounts of fractured soft rock such as shale, siltstone, sandstone, and especially the light diatomaceous and cretaceous formations.

It is often thought that crawler-type grading equipment, in view of its great weight, can also be used with success for fill compaction. However, the main purpose of the crawler tracks is to spread this weight over a large area so that the equipment can move over very soft terrain. The pressure on the soil below the tracks is usually less than 100 psf, which is less than half that applied by a man standing on both feet. The compactive effort of such a low pressure is insignificant. Rubber-tired grading equipment is of more use in this respect because the tire pressures are relatively high and the tire sizes can be very large. To be effective, the equipment must take a different route on every trip, a procedure which is not very easily followed on most projects. There is a high probability that narrow strips of soil will be compacted very well, leaving ridges of uncompacted soil between them.

A typical grading and compaction project with several types of equipment at work is shown in Fig. 11–3.

FIG. 11–3 *Small grading project. Field density test is performed in left corner.*

Compaction in Layers.   All compaction of soil and rock fill must be done on successive layers of the material each of which is subjected to enough compactive effort to obtain the desired degree of density. For a given type of equipment and soil, there is an optimum combination of layer thickness and number of passes of the compactor which will give the desired degree of compaction with the minimum total compactive effort or greatest economy. There is also a maximum layer thickness that can be compacted satisfactorily with a certain piece of compaction equipment no matter how many passes are made. Usually the layer or lift thickness selected is close to this maximum, varying from 3 in. for small rollers and vibrating plates to 9 or even 12 in. for very large rubber-tired rollers in cohesionless soil. The number of passes required for the specified degree of compaction can be determined by trial. Unit weight determinations made in the compacted soil should not be confined to the upper few inches of a lift, but should extend into the lower part, where the chances of inadequate compaction are greatest. The number of passes required is usually found to be below 10; a greater number of passes gives proportionally less increase of density.

Specifications and Control.   The foregoing brief summary of compaction methods shows the wide variety of equipment available to obtain the desired degree of compaction of a soil. Established practices have been subject to constant change and improvement; soils which a few decades ago would have been rejected as unsuitable for compaction are now used and compacted without difficulty with newly developed equipment. The engineer who must prepare specifications for earth movement and compaction should seriously consider the question of whether to specify the method of compaction or the desired condition of the compacted fill. By specifying the method of compaction the contractor is limited in his choice of equipment, the lift thickness, and the number of passes for each lift. Notwithstanding the fact that the engineer may have a wealth of experience in fill compaction, the possibility exists that a more efficient or economical method of compaction than the one specified could produce the desired result. In practically all cases it will be preferable to specify the condition of the finished fill, rather than the method of compaction, in accordance with common practice in the other phases of construction.

The most important property to be specified is the degree of density or the dry unit weight of the compacted fill. For this purpose a number of standardized laboratory compaction methods have been developed; practically all of them consist of compacting the fill with a drophammer in a 4- to 6-in.-diam mold. In the United States the most frequently used laboratory compaction tests are those specified by the American Association of State Highway Officials (AASHO) and the American Society of

Testing Materials (ASTM). As an example, the ASTM D 1557–64 T procedure, Method C, is described briefly below.

The compaction mold is shown in Fig. 11–4. The inside diameter is 4 in., and the volume of the lower part of the mold is $\frac{1}{30}$ ft^3. The rammer, which may be manually or mechanically operated, has a 2-in.-diam face and weighs 10 lb. The sample is thoroughly mixed and passed through a $\frac{3}{4}$-in. sieve. Only the material passing the sieve is used in the test. At least 12 lb of material is needed for convenient work. The sample is mixed with water to give it a moisture content somewhat below the optimum for compaction. (This is a matter of judgment, and an experienced operator can obtain a curve such as that shown in Fig. 11–1 in as few as three well-spaced points.) The sample is compacted in the mold in five layers of a compacted thickness of about one inch each by 25 uniformly distributed blows from the rammer falling from a height of 18 in. Following compaction, the extension ring is removed from the collar, the sample is trimmed off and weighed, and a small sample is dried for moisture determination. The test is repeated until the peak value has been passed.

Other tests vary mainly in the compaction energy, mold size, or size of coarsest soil fraction. A test sometimes used for fine soils is the Harvard miniature compaction test, which uses a small ($\frac{15}{16}$-in.-diam) mold and a spring-loaded tamper. The apparatus size is so designed that the weight of the sample in grams is equivalent to the wet unit weight in pounds per cubic foot. None of the laboratory tests imitates exactly the method of compaction by kneading and rolling which takes place in the field, but the tests do establish a dry unit weight standard which can be used in the specifications. Usually the specified dry unit weight is 90 or 95 percent of that obtained with one of the laboratory compaction tests mentioned.

In many cases, other properties in addition to the degree of compaction must be mentioned in the specifications. If the source of borrow material is not known at the time the specifications are written, the type of fill suitable for the project must be specified. Narrow limits for grain-size distributions are required for dam fills and non-frost-susceptible highway fills. To obtain greater strength in a compacted fill it is often specified that

Fig. 11–4   *Standard 4-in.-diam compaction mold.* (Courtesy Soiltest Inc.)

the soil be compacted at a moisture content slightly below the optimum moisture content as determined from the laboratory compaction tests, whereas, on the other hand, to obtain greater flexibility or lower permeability a higher moisture content can be specified. The maximum allowable size of rocks in a fill is usually taken as somewhat less than the lift thickness, and specifications exclude all organic material and topsoil. Furthermore, specifications cover requirements for the preparation of a site before a fill can be placed; these will include, for example, removing vegetation and other undesirable material and eliminating wet surface conditions by first placing a blanket of coarse granular fill.

Field control of fill compaction is of great importance for maintaining the same quality of work regardless of weather conditions or changes in borrow material. This is facilitated by clear specifications which leave the contractor no doubt about the standards of work expected. The control is exercised through the results of field tests of unit weight and moisture content of the fill. To determine the field unit weight of fills, a number of tests are in use; most of them are based on the simple steps of measuring the volume of a small hole excavated in the fill and determining the weight and the moisture content of the material removed from the hole. The volume of the hole is measured, for example, by filling it with a uniform sand the unit weight of which is known to vary between very narrow limits when poured loosely into a hole, or by expanding a water-filled balloon into the hole. In cohesive fills a very rapid method of volume determination is submergence of a lump of the soil in oil or mercury. A typical test for in-place density determination of soils, ASTM Designation D 1556–64, is described below.

The density apparatus (Fig. 11–5) consists of a baseplate, a metal funnel, and a gallon jar. The jar is filled with a clean, dry, free-flowing sand of fairly uniform grain size. The unit weight of this sand, when poured loosely into a container, may not vary more than 1 percent. The apparatus is calibrated by placing the cone on a level surface, opening the valve between jar and cone, and measuring the weight (and thus the volume) of the sand which fills the cone and the valve pipe. A hole with the same diameter as the cone is dug by hand; the excavated material is weighed immediately; and a small sample is preserved in a moisture-tight container for moisture determination. The baseplate is seated firmly over the hole; the cone is placed on the baseplate; the valve is opened, and the hole, cone and valve stem are filled with sand. The valve is closed, and the amount of sand lost is determined from the weight of the apparatus before and after the test. This provides the volume of the hole, and the unit weight of the fill can be determined.

The field portion of the test provides only the wet unit weight of the fill, and the moisture content must be determined before the dry unit weight can be calculated. The drying of the soil sample takes considerable time;

Fig. 11–5   *Sand-cone method of field density determination.* (Courtesy Soiltest Inc.)

more than can be afforded at today's pace of grading operations. Rapid methods for moisture content determinations have therefore been developed; they feature radiant heat and alcohol, carbide, and other hygroscopic admixtures for the rapid removal and measurement of soil moisture. An experienced field inspector will have little trouble estimating the moisture content of a fill after having worked with it for a few days, and his judgment can generally be used with confidence pending the outcome of the moisture content determinations which must be made for the records.

Since larger stones are removed from the fill for the laboratory compaction tests, the effect of the stone content on the field density must be taken into account. Correction factors have been developed for this condition, but their use involves making frequent grain-size analyses in the field, which is again a time-consuming operation. A more practical approach has been to exclude the stones from the volume and weight measurement of the soil in the field, thus confining the control figures to the soil matrix. A very important condition for successful and fair field control is that a large enough number of tests is taken at various levels in each lift to enable a reasonable average value of unit weight and moisture content to be determined.

As mentioned in Chapter 7, radiational equipment for density and moisture determination is available, and it can perform a large number of field tests in a short time. The economy of the equipment, taking into account the repeated calibrations required and the high cost of the apparatus, is still somewhat in doubt in many applications.

Engineering Properties of Compacted Soil.   As mentioned above, compaction of a soil offers the opportunity of manufacturing the soil in accordance with specific needs regarding strength, compressibility, flexibility, permeability, etc. This is done premeditatedly in pavement design and in the design of earth dams. Samples of the fill available for such projects are compacted and tested in the laboratory to determine the properties of importance to the design. Since fill containing coarse gravel or other large fragments cannot be tested readily in the small samples normally used in the laboratory, a test fill may be placed for experimental purposes.

With modern compaction equipment and adequate control, practically all soils can be compacted to a state of density well in excess of that in their natural condition, thereby creating a strong, stable mass of low compressibility and porosity. Quality control can be at least as effective as, for instance, for site-mixed concrete. In a 250-ft-high earth-fill dam, the lower part is under a static pressure of 20,000 psf and over; yet when the reservoir is filled and the compacted fill gradually becomes saturated, the subsequent settlement is usually of the order of 1 percent of the height or less. Although settlement of this magnitude is still regarded with some regret by the dam designers, it compares favorably with the performance of most natural deposits. In foundation engineering, allowable footing pressures of 4,000 psf and more are feasible in most well-compacted soils, considering both stability and settlement. Naturally, the higher the demands that are made on a fill, the more exacting the compaction control should be.

A question arises regarding the ability of a compacted fill to retain its favorable properties with time: is there a tendency toward swelling and gradual deterioration of the artifically stabilized soil? The answer for coarse granular, noncohesive soils is no. Only high pore pressures induced by impact, seismic, or vibratory loading under saturation, or high seepage forces due to groundwater flow could cause a rearrangement of the soil particles and loosening of the compacted structure. In cohesive soils, however, partial loss of strength and increased compressibility may be caused by an increase in moisture content. It has been found[1] that cohesive soils compacted at a moisture content below the optimum moisture content for compaction are very strong but lose a considerable amount of this strength upon saturation, particularly if there is little confinement of the fill. Clays compacted at a moisture content above optimum have a lower strength than those compacted to the same dry unit weight but on the dry side of optimum, and little additional loss of strength takes place upon saturation. If the possibility exists that a fill will become fully or partially saturated during its lifetime, the total design should be based on the properties in a saturated condition, and consideration should be given to specifying compaction on the wet side of optimum. On the other hand, water contents on the wet side of optimum may be great enough to permit substantial pore pressures to develop in the fill mass during construction. The presence of these pressures and their dissipation have an effect on the stability and settlement of the fill.

Well-graded sand-gravel mixtures or fine to coarse sand without gravel can be compacted to a very dense, competent foundation material which will be affected very little by subsequent saturation. However, the compactive effort is largely wasted on sands which are nearly uniform in grain size. The difference between maximum and minimum density for these soils is very small, and any form of disturbance or vibration may undo the work of the compactor.

11-3 HYDRAULIC FILLS

In addition to the common methods of earth movement with grading and compaction equipment, fills are also placed by mixing the soil with water and depositing the resulting slurry at the desired location either on land or at the bottom of a creek, lake, or bay where the soil particles settle out to form the structure. This is called a hydraulic fill operation, and in most cases the fill is obtained from below water by using a suction dredge. As can be imagined, clay particles will remain in suspension too long to be used in hydraulic fill, and most hydraulic fills consist of sand and silt only. The area to be raised with a hydraulic fill is first surrounded with a low dike or levee. As the area fills up, the excess water will spill over the levees, drain off, or be absorbed into the earth. Sand deposited in this manner will have a fairly high relative density, offering good foundation support. Silt tends to form a more porous structure which will continue to consolidate under its own weight for a considerable time, and this presents difficulties for future construction. The rate of consolidation can be accelerated with sand drains, as was described in Chapter 10.

Hydraulic fills placed under water are frequently found unsuitable for foundation support not because of the condition of the fill, but because of the soft mud and young alluvium which underlie the fill. In poorly controlled fills the soft mud is often partially displaced by fill, leaving localized areas where the mud is very thick and the fill cover very thin. In many cases it would be more economical to dredge and dispose of the mud before placing fill. The additional cost of this operation may compare favorably with the savings obtained from being able to found substantial structures directly on the fill material.

The same method of deposition which is so successful in sandy hydraulic fills is used in the construction industry to compact sand backfill behind retaining walls and even to compact sand fill under floors. The sand is dumped without effort to compact it and is subsequently flooded and jetted to a higher degree of density. For fills under floor slabs this method is not very effective, owing to the relatively small seepage forces developed. In backfill behind retaining walls with good underdrains excellent results can be obtained, particularly if the sand is agitated with a vibrating needle at the same time.

11-4 SOIL STABILIZATION WITH ADDITIVES

Improvement of soil properties by mixing and compacting the soil with other substances such as cement, bituminous mixtures, or chemicals has long been an established practice in highway engineering. More recently, chemical stabilization of in-place soil has gained popularity in the construction industry, in particular for the support of existing foundations exposed by adjoining excavation, tunnel construction, and open excavations. The most common additives and their influence on the soil will be discussed.

Cement. Cement is used in three forms of soil stabilization: compacted soil-cement, poured soil-cement mixtures, and cement grouting. The term "soil-cement" has become very familiar in highway construction, and thousands of miles of soil-cement roads have been constructed and are in use. The technique is also used in airfield construction and for slope protection and canal and reservoir linings.

The action of cement in soil-cement mixtures is the same as in concrete: cement hydration binds soil particles together. The soil properties are affected to a degree depending on the amount of cement used. For instance, low percentages (below 2 percent) of cement by weight used in heavy clays have little effect on the strength but tend to reduce the plasticity, and the term *cement-modified soil* is often used for such mixtures. Higher percentages of cement will begin to cause an increase in strength of the compacted soil-cement mixture. As with concrete, the hydration of the cement will be affected adversely by organic substances in the soil or the presence of dirty or oily water or salt water.

The chief factors determining the performance of a soil-cement are soil type, moisture content, method and uniformity of mixing, and compaction. Practically any nonorganic soil can be improved or stabilized by the addition of cement. The effect of the cement is usually determined from simple compression tests on cylindrical specimens of compacted, cured soil-cement with a height equal to twice the diameter. For soil-cement mixtures with a cement content of 10 percent by weight, the compressive strength may vary from 100 psi in silts and poorly graded sands to as high as 1,000 psi in well-graded gravel-sand-clay mixtures. For comparison, a well-graded compacted gravel-sand-clay mixture without cement would rarely have a compressive strength in excess of 50 psi.

In general, the higher the cement content, the greater will be the strength of the soil-cement. The rate of increase in strength decreases somewhat for high percentages, but the main limit to cement content is cost. For satisfactory performance the cement content will be from 5 to 12 percent by weight for sandy and gravelly soils and from 12 to 20 percent for silt and clay. High-early-strength cement appears to give better results than ordinary cement.[2] Although laboratory results usually express the cement content as a percent of dry weight, the field procedure is to specify cement content as a percent of volume of compacted soil cement. A 3 percent cement content by weight corresponds roughly to one sack (94 lb) of cement per cubic yard of compacted soil-cement.

Moisture content affects the strength of soil-cement in the same way that it affects the unit weight of a compacted soil. The optimum moisture content for compaction of a soil changes very little with increasing cement content. The curves showing the dry unit weight–moisture content and the compressive strength–moisture content relationships are similar in shape to the dry unit weight–moisture content curves shown in Fig. 11–1 for plain soil. The maximum strength is obtained at the maximum unit weight, which is reached at a moisture content very close to the optimum.

The effect of the quantity of cement used in the field depends on the adequacy with which it is mixed with the soil. Good, uniform mixing will achieve the same results with less cement than must be used with poor or little mixing. Two methods of mixing soil-cement are possible: mixing with a traveling plant and mixing with a stationary plant. In the latter case, the soil is excavated and taken to the plant, where water and cement are added. After mixing, the mixture must be hauled back to the site, spread, and compacted. The maximum allowable time between mixing and compacting is usually specified to avoid substantial hydration before the soil-cement is compacted. This time limit as a rule is several hours and does not seriously hamper a well-organized operation with a stationary plant. A mobile plant has the advantage of eliminating excavation and transportation, but the equipment is very elaborate and requires substantial support. The quality control and thoroughness of mobile mixing are less than can be attained with a stationary plant. A traveling plant may either disturb and mix the soil in place in the same way as does a rotary tiller or lift the soil and mix it inside the plant.

Compaction is just as important for soil-cement as for plain soil. Pneumatic-tire rollers, smooth-wheel rollers, and vibrating rollers are used in the same manner, with the total thickness of compacted soil-cement usually being 6 to 8 in. With a traveling plant this must be compacted in one lift. Field control is carried out by unit weight determinations as with normally compacted soil and is supplemental to strength tests, which will be described under Pavement Design later in this chapter. The as-mixed cement content cannot be determined easily in the field and should be controlled at the plant.

Lambe and co-workers[3] have described research on the effect of certain additional chemicals on the performance of soil-cement. Under laboratory conditions the cement stabilization of some organic soils and reduction of the cement content in other soils were found possible upon addition of small amounts of sodium sulfate, calcium chloride, and alkali-metal compounds.

Other forms of soil stabilization employ soil-cement in plastic, poured soil-cement, or cement grout forms. Plastic soil-cement mixtures are sometimes used in reservoirs and channel linings. The cement content is higher than in compacted soil-cement, and the possibility of shrinkage is greater. The degree of shrinkage depends mainly on the soil type and must be investigated as part of a feasibility study for this type of lining in connection with a given structure. Cement grout is used mainly for grout curtain walls in fractured and porous rock formations below dams. A cement-water slurry is pumped under high pressure into a row of drilled holes in an attempt to form a continuous, impervious cutoff wall below the dam.

Bituminous Stabilization. Bituminous material used for soil stabilization is usually asphalt with a relatively low softening temperature. The viscosity of the asphalt can be reduced by emulsification in water or by addition of gasoline or other solvents (the product is referred to as cutback asphalt). Nearly every nonorganic soil can be stabilized by the addition of 2 to 10 percent asphalt by weight, but bituminous admixtures are not very effective in very coarse soil (50 percent or more gravel) or in very plastic clay (liquid limit over 50 percent). The effect of bituminous admixtures is to bind the soil particles together and to waterproof soils which normally soften when in contact with water.

In many respects, bituminous soil stabilization is comparable with cement stabilization. The more asphalt used, the greater the strength which can be obtained, up to the limits of workability of the mixture. This limit is lower than with cement, and more than 10 percent asphalt by weight is seldom used. The addition of asphalt decreases the maximum dry unit weight and increases the strength, particularly the strength after saturation. The moisture content–dry unit weight relationship is similar to that for compacted soils and soil-cement mixtures, but it is customary and more practical to include all volatiles and to plot the dry unit weight versus the volatile content. Here, however, the problem that arises is that the content of volatiles required for good mixing is greater than that required for good compaction, and the volatile content required for good compaction in turn is much greater than that required for optimum stability. As a result, the stabilization process, which is basically the same as for cement stabilization, must be interrupted for drying before compaction and again before application of the surface finish. After curing in this manner, the same range of compressive strengths can be attained as was given for compacted soil-cement. The equipment used for bituminous stabilization is practically the same as that used with soil-cement. Control in the field is somewhat more complicated, and the mix design requires many laboratory trials.

Lime Stabilization. Lime mainly used in soil stabilization is hydrated lime [$Ca(OH)_2$], which may contain minor amounts of MgO or $Mg(OH)_2$. The addition of lime to wet soil initially improves the workability by increasing the plasticity of soils with low plasticity and decreasing the plasticity of highly plastic clays. After a curing period, cementation sets in, thereby increasing the strength of the compacted mixture. The resulting compressive strength is somewhat lower than that which can be obtained with stabilization with cement or asphalt, but it still represents an appreciable increase in strength over that of the compacted untreated soil. The optimum moisture content for compaction increases with increasing lime content, and the dry unit weight decreases simultaneously. The lime content required for effective treatment is about 5 percent of dry weight for granular soils and up to 10 percent for cohesive soils. The field procedure is again quite similar to that for soil-cement and bituminous stabilization, and the same equipment can be used. A curing period of about 5 days is required for adequate cementation.

Chemical Stabilization. Other chemicals have been used occasionally for soil stabilization in road construction and for lowering the susceptibility to frost damage of some soils, but none of them are of enough importance to be discussed at length in this book. Another form of stabilization with chemicals is gaining prominence, namely, stabilization of in-place soils by spraying or grouting with chemicals. Spraying is used only for surface treatment of excavation sides and reservoirs; it provides a very thin cover of stabilized soil. Injection is the more frequently used procedure for chemical stabilization, particularly of noncohesive soil with a permeability sufficiently high to make the soil receptive to grouting. Most chemicals used in this form of stabilization consist of a solution of sodium silicate in water (water glass) plus a catalyst; the combination of the two solutions in the pores of the soil causes formation of an insoluble silica gel. Because of the rapidity of the chemical reaction, most systems use what is called "two-shot grouting," injecting the chemicals in turn. The strength of the stabilized mass can be very high, and spectacular feats of temporary underpinning and excavation stabilization have been accomplished with the aid of the method (Fig. 11–6). To slow down the reaction, additives are frequently employed; mostly such additives are patented. The durability of the stabilization effect cannot always be assured.

Another chemical grout used widely is known by the trade name AM-9.

FIG. 11−6 *Chemically stabilized soil in excavation below existing building.* (Courtesy Spencer Soil Solidification)

This chemical has a very low viscosity and can be injected in a single shot. It can, however, be harmful to personnel and must be handled with caution.

11−5 PAVEMENT DESIGN

All soil engineering concerns itself with the attempt to make the soil in a given area accept the loads which people wish to impose upon it. The loads take a variety of forms above the surface, but the end result is that a certain stress distribution in the underlying material may not exceed the strength characteristics of that material in whatever way they are defined. The problem of designing a highway or airfield pavement to distribute a given automobile, truck, or aircraft wheel load into the underlying soil differs only slightly from the problem of imposing a column load at the soil surface. The difference, and it is one which is rarely taken into account in any numerical way in practice, is due to the fact that a wheel load is transient. Therefore, transient stresses or vibrations which may result from impact loadings are transmitted into the soil, although static loads are possible. Since these transient loads will, in many cases, be repeated indefinitely, some considerations of a fatigue type of failure in the soil underneath a highway or airfield may be expected. A highway, being a relatively narrow track along which vehicles are constrained to run, transmits loads to the same areas of underlying soil with a much greater frequency than is the case with an airfield. Aircraft are relatively free to land at different locations and may take different paths down the runway to a greater extent than the average road vehicle is free to do. On the other hand, the loads borne by a highway pavement are much smaller in magnitude than those to which the airfield is subjected.

Initially, roads consisted merely of a cleared stretch of ground surfaced only with the natural soil or, on rare occasions and perhaps in cities, with paving stones, while the early airfields consisted merely of grass fields, as their name implies. Greater traffic on the roads led to deterioration of the roads, particularly in areas of poor-bearing-capacity soil, and the necessity of developing some sort of wearing surface arose. In the case of airfields, the growing maximum weight of aircraft and the development of retractable undercarriage with necessarily smaller wheels brought with them the necessity for designing the airfield surface to take greater loads concentrated in smaller areas, since there was a limit to the bearing capabilities of grass fields underlain by different types of soil.

It was of no value to place a wearing surface or a road on top of a soil which was unable to support the wheel loads, and consequently from the earliest times some form of artificially prepared base course was placed on top of the natural ground; this was overlaid with the wearing surface to take the vehicles. Thus, the necessity arose for deciding in any given circumstances the material to be employed in the base course and the thickness which must be placed on top of a soil. The layers of material then are intended to distribute the applied stresses in such a way that the strength of each layer is not exceeded.

Pavement construction is divided generally into two types, the so-called flexible and rigid pavements. The former consist of a surfacing of bituminous or asphaltic material placed on top of the base course, and in the latter a portland cement concrete surface is used. In certain cases of underlying soil with high bearing capacity, the concrete may be placed directly on top of the surface. The distinguishing characteristics of the asphaltic surface are its low tensile strength and its relatively low compressive strength. On the other hand, portland cement concrete has considerable tensile strength (especially when reinforced with steel bar or mesh) and compressive strength, and these properties may be utilized in the design of surfaced roads.

Flexible Pavement Design. The same categories of design methods found elsewhere in soil engineering are applied to pavement design also. The methods may be classed as (1) wholly empirical using no soil-strength tests, (2) wholly empirical using some form of soil-strength tests, (3) partly empirical, partly theoretical using strength tests, and (4) entirely theoretical using strength tests. Of these methods, the fourth can be disposed of immediately, since there are no entirely theoretical methods of designing pavements of either type.

Since the base course material is most frequently a gravel fill meeting some standards of durability, the normal requirement of any design method is to establish for a particular soil the thickness of construction that is necessary to support given magnitudes and frequency of loadings satisfactorily; the thickness of construction normally includes the surface course. As a consequence of the study and correlation of road construction and performance data, the construction thickness for a given loading has been found to be related to the underlying soil type. The identification of the soil is carried out with reference to one of the classification systems, such as that described in Table 1–2. A design method such as this may be designated "wholly empirical." To accomplish the necessary identification of the natural soil, only rather simple soil tests are required, and they involve no direct measurement of shearing strength or the soil's behavior under different stresses. Besides the tactual and visual identification procedures outlined in Chapter 1, the soil can be readily classified by a sieve and hydrometer analysis together with the liquid and plastic limits of that portion of the soil passing the No. 200 sieve.

As an example of a wholly empirical design technique, the following method may be cited. The U.S. Bureau of Public Roads has set up an index, called the soil group index, which may be represented by the following equation:

$$G = 0.2a + 0.005ac + 0.01bd \tag{11-1}$$

where a = that portion of the percentage passing a No. 200 sieve greater than 35 and not exceeding 75, expressed as a positive whole number

b = that portion of the percentage passing a No. 200 sieve greater than 15 and not exceeding 55, expressed as a positive whole number

c = that portion of the numerical liquid limit greater than 40 and not exceeding 60, expressed as a positive whole number

d = that portion of the numerical plasticity index greater than 10 and not exceeding 30, expressed as a positive whole number

When a particular soil has been analyzed and a group index determined according to Eq. (11–1), a table of values or a chart such as the one given in Fig. 11–7 can be used to give the required thickness of construction and the thickness of base course for a given type of load. No exhaustive analysis of frequencies and vehicle weights has been attempted. The loads are merely classified according to light, medium, and heavy traffic, for which three curves are given. Local road departments have modified the curves to accommodate special soil conditions. It appears, in general, that a given road is safe against failure when the group index method is used, but no doubt the thickness is greater than necessary in many cases.

Certain deficiencies are obviously inherent in the group index method, since the shear-strength capabilities of a given soil are only partially accounted for by its position in the classification table and soil strength may vary according to water content, previous stress history, and other factors. Additionally, the actual loads and frequency of loading are not expressed in the method from the point of view of the stresses imposed on the underlying soil. The growing dissatisfaction with such purely empirical techniques as the group index method led, in the 1930s, to attempts at the still empirical classification of soils but one involving some kind of strength test of the subgrade material. It seemed logical to test the soil by a device duplicating, if possible, the manner in which the stress was imposed on the soil by a wheel load on the pavement surface. Thus, varieties of plate bearing or penetration tests were introduced, and these tests yield curves of penetration distances or settlement amounts versus load that are characteristic of the strength of a soil. Perhaps the best known of these methods of indirect empirical determination of the shearing strength of soil was developed by the California Division of Highways under O. J. Porter and is known as the California bearing ratio method or CBR method.

It would be desirable, in a test of this type, to subject the soil to a load whose contact area corresponded, at least roughly, to that which is to be stressed by future traffic or other loads. This would result in an uneconom-

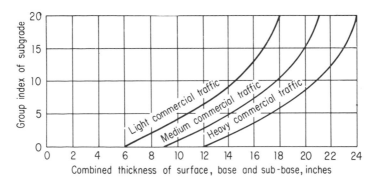

FIG. 11–7 *Typical design curves, U.S. Bureau of Public Roads group index method.* (After Steele)

ical test because of the size of plate and dead load required. In addition, it is preferable that any test of the soil strength should also be paralleled by a similar laboratory determination. For these reasons, the California Division of Highways adopted a simple penetration test in which the force required to push a small piston or plunger 3 in.[2] in area into the soil at a given rate is measured, along with the penetration of the plunger. With such equipment, loads greater than about 3,000 lb are not required, and inexpensive and rapid field and laboratory tests can both be performed.

In the CBR test as presently used, the rate of piston penetration has been standardized at 0.05 in./min, and a curve is plotted of load versus penetration as shown in Fig. 11–8. For a given soil, the load at 0.1 or 0.2 in. of penetration is then divided by a standard load of 3,000 or 4,500 lb, respectively. The proportion of the measured load to the standard load as a percentage is taken to be the CBR for that particular soil under the test conditions. The standard loads are based on those obtained in a dense compact gravel, so that the higher the CBR value, the better is the soil as a base for a pavement.

The CBR test was subsequently revised, extended, and standardized by the U.S. Army Corps of Engineers during World War II for the purpose of devising airfield flexible pavement design procedures. In the CBR method as developed by the Corps of Engineers, samples of any compacted soil which may be utilized in construction and subjected to wheel loadings are prepared in standard CBR molds in the laboratory and compacted to the same density as will be specified for field control. These samples are then subjected to 4 days of soaking under water (to simulate the possibly harsher field moisture conditions to which the actual construction might be subjected) before the CBR test is carried out. In this respect the British Division of Building Research feels that the soaking of the specimens results in an uneven distribution of moisture content and swelling at the ends of the specimen to such an extent that the ends where the CBR test is carried out are softened and a misleading low value of CBR is obtained. This method has been considered to yield an extremely conservative design of pavement.

The California Division of Highways and later the U.S. Army Corps of

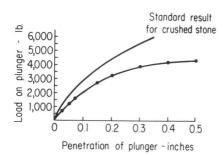

FIG. 11–8   *Typical CBR test results.*

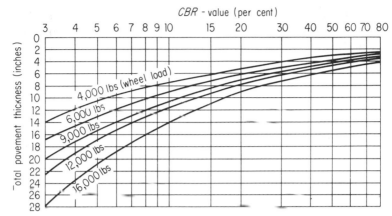

FIG. 11-9   *Corps of Engineers' highway design curves for flexible pavements.*

Engineers carried out tests in which the loads, construction thicknesses, and CBR values of both the natural soil and base course material were varied for the purpose of preparing design curves. For different wheel loads these then give the thicknesses of construction which are required over any subgrade with a given CBR value. The CBR of the base course is required to be 80 or greater. The curves used by this method are shown in Fig. 11-9, the use of which is self-evident.

More recently, the California Division of Highways has replaced the CBR laboratory test with a different type of test in an apparatus called the Hveem Stabilometer. The apparatus is somewhat similar to the triaxial cell referred to in Section 5-1. A cylindrical sample 4 in. in diameter and 2.5 in. high is subjected to a cell pressure up to 100 psi and compressed axially while the deformation is measured. The deformation is related to the required pavement thickness in the same empirical manner as with the CBR test. Since the success of empirical tests of this nature depends directly on the statistical evidence relating test results to actual field performance, the CBR test, after more than thirty years of application, remains the preferred method.

Turning to design techniques involving theory, it is noted that in few of the methods that have been used has a theoretical stress distribution in the soil subjected to a surface wheel load been compared with the actual strength of the soil determined from some type of test. The logic of this method is apparent, but practical objections to its application are equally obvious. Ideally, a given load is applied to the surface of a pavement over a certain area; both the load and the area can be relatively closely specified in practice. Underneath the pavement surface lies the base course whose elastic constants may be assumed known, as are those of the pavement. Below this is the natural subgrade whose physical properties can, in principle, be determined by testing. As a first approximation the physical properties of the pavement surface, base course, and subgrade can be assumed to be the same, and a stress distribution theory such as due to Boussinesq or Westergaard may be applied to give the variation of maximum shear stress, say, underneath the load. The construction thickness is so chosen that the shearing strength of the soil as indicated by tests everywhere exceeds by some chosen margin this shearing stress. The difficulties of the application of this method are many, and they may be briefly summarized.

In such a technique, the first approximation of a homogeneous material will almost never be a practical one, and great stress variations accompany the presence of layers of differing properties. These properties are very difficult to determine; moreover, the equations for the determination of stresses in a layered material are difficult to evaluate. As a last point, the assumption that failure will occur when a shearing stress exceeds a shearing strength is unlikely to lead to a realistic failure mechanism. The difficulties of evaluating the shearing strength of a soil have been pointed out before.

Golder[4] carried out an investigation of various roads both in satisfactory and in failed conditions in which he examined the variation of shearing strength with depth, and he determined what distribution of stress separated the failures from the successful performances. From his results he concluded that the construction thickness necessary to support a given load was one that caused the shearing stress everywhere in the subgrade, as calculated from Boussinesq's equations for a homogeneous profile, to be less than the shearing strength of the soil as measured by an unconfined compression test.

Rigid Pavements.   In many cases, the design of rigid-slab construction follows simply along the lines of, say, the CBR method as outlined above for the flexible pavement except that the rigid slab is included in the thickness of construction with no allowance being made for its rigidity. However, the more advanced design of rigid pavements for heavier loads has certain features which distinguish it from flexible pavement design, since the strength of the pavement surface slab itself must be taken into account and it is necessary to design the slab economically. In order to design a rigid pavement the stresses that may be induced in the slab by means of the wheel loading must be known, and these stresses must not exceed the strength of the slab in any one of its several failure modes.

In the early days of aircraft and highway vehicle design, only single wheels were used at the points of support of the airplane or automobile, and methods were derived for calculating the stresses in pavements underneath the single loads on small areas at the surface. With the growing weight of aircraft and size of transportation on highways it became common to use two wheels in tandem or abreast (dual) to support the increasing loads. Along with this development came the design of aircraft with retractable undercarriages, so that smaller wheels which would fit into the available space in wings or fuselage became necessary; higher tire pressures followed. Again, with even greater weights of aircraft it became necessary in some cases to provide not only dual or tandem wheels for each undercarriage leg but also systems composed of dual-tandem arrangements of wheels. The problem that therefore arises is what stresses are caused in the pavement, base course, and subgrade as a result of two or more closely spaced wheels. The method adopted is to analyze the stress resulting from various wheel arrangements and to calculate what single wheel load would produce the same stress. This is then known as the *equivalent single-wheel load* (ESL), and for any arrangement of tires, tables which show the ESL are available. The stresses due to single wheels have been analyzed, so that the reduction of the problem to the loading of a single wheel enables the stresses to be evaluated.

The stresses in a reinforced-concrete pavement resting on another material, due to a single-wheel load at the ground surface, have been given by Westergaard.[5] The usual reinforced-concrete slab construction technique involves pouring the slabs in rectangles some 15 to 20 ft by 10 or 12 ft to allow for expansion or contraction of the slab. Therefore, the worst conditions of loading in the concrete may result when the load is applied at the corner or at the edge of a slab. The critical stresses in concrete are tensile, and equations for these have been given by Westergaard for a single-wheel load acting at the edge or corner of a concrete slab resting on another material:

$$\sigma_c = \frac{3P}{h^2} \left\{ 1 - \left[\frac{12(1 - \mu^2)k}{Eh^3} \right]^{0.15} (a \sqrt{2})^{0.6} \right\} \qquad (11\text{--}2)$$

$$\sigma_e = \frac{0.529P}{h^2} (1 + 0.54\mu) \left(\log_{10} \frac{Eh^3}{kb^4} - 0.71 \right) \qquad (11\text{--}3)$$

$$\sigma_i = \frac{0.275P}{h^2} (1 + \mu) \log_{10} \frac{Eh^3}{kb^4} \qquad (11\text{--}4)$$

where E = modulus of elasticity of concrete, psi
μ = Poisson's ratio for concrete
k = modulus of subgrade reaction, psi/in.
h = thickness of concrete slab, in.
P = total load exerted by one wheel, lb
a = radius of equivalent circle of contact area, in.
b = $\sqrt{1.6a^2 + h^2} - 0.675h$ if $a < 1.274h$; $b = a$ if $a > 1.274h$
σ_c = maximum tensile stress in concrete at the top of slab near corner, psi
σ_e = maximum tensile stress in concrete at bottom of slab along unbroken edge, psi
σ_i = maximum tensile stress in concrete at bottom of slab at an interior location, directly under center of applied load, psi

The equations indicate that the worst case occurs when the load is applied at the center of one edge of the slab. The stress varies directly with the magnitude of the load and indirectly with the thickness of the slab to the nth power, so that stresses can be very quickly reduced by thickening the slab. Analyses of the stress distribution in concrete slabs with various temperature variations through the slab thickness have been made but will not be given here.

In almost all pavements, some form of steel reinforcing is used. In highways steel mesh reinforcing is used for the prevention of cracking from temperature variations; in airfields No. 10 or No. 12 bars at 6- or 8-in. centers in two directions are used, unless extremely heavy aircraft use the airfield, in which case even more reinforcing is used. As a result of the Westergaard equations which indicate greatest tensile stress when a load is applied at the edge of a slab, it has been common design practice in the United States to thicken the edges of both highway and airfield pavements.

In the center of the runway or highway where a wheel may roll from one slab to the next, some method of transferring the load from slab to slab to provide continuity must be adopted. Occasionally, the concrete slabs are keyed to one another, but a method of growing importance is the use of dowel bar construction. Here, during construction, steel rods are laid in the edge of one slab at fixed intervals, depending on the load, and allowed to project while the slab is being poured. The projecting end of the dowel bar is covered with a tube sheath lubricated on the inside, and the adjacent slab is poured. In this way the dowel bar is free to expand or contract, or the slab itself may move longitudinally without developing longitudinal stresses in the bar, but the bar is able to transfer a vertical load from one slab to another. It has been found that the alignment of the bars is very critical, and therefore precise construction methods should be used to ensure that the bars are truly parallel and horizontal.

So far we have considered the stresses from the point of view of the slab alone, but as noted before, a constant k, the modulus of subgrade reaction, occurs in the equations. The constant represents the effect of the underlying soil on the pavement slab in a manner similar to that in which the soil below a continuous footing affects the behavior of the footing and the stresses which are induced in it. Westergaard, in his analysis, gave no indication of the method by which k was to be measured in the field, but tests of some form are obviously necessary in order to determine its magnitude. Enough of these tests have been carried out to date to enable a correlation to be made between the value of k for any particular soil and either the CBR for the material or a rough identification of the soil type according to the UCS. In Table 11-1 the values of the modulus of subgrade reaction and CBR are plotted on the UCS in such a way as to indicate the magnitude of the modulus for any given soil type. It must be

TABLE 11-1 CORRELATION BETWEEN UCS, FIELD CBR, AND MODULUS OF SUBGRADE REACTION (CORPS OF ENGINEERS)

UCS symbol	*Soil description*	*Field CBR, %*	*Modulus of subgrade reaction, psi/in.*
GW	Gravel or sandy gravel, well graded	60–80	300 or more
GP	Gravel or sandy gravel, poorly graded	35–60	300 or more
GU	Gravel or sandy gravel, uniformly graded	25–50	300 or more
GM	Silty gravel or silty sandy gravel	40–80	300 or more
GC	Clayey gravel or clayey sandy gravel	20–40	200–300
SW	Sand or gravelly sand, well graded	20–40	200–300
SP	Sand or gravelly sand, poorly graded	15 25	200–300
SU	Sand or gravelly sand, uniformly graded	10–20	200–300
SM	Silty sand or silty gravelly sand	20–40	200–300
SC	Clayey sand or clayey gravelly sand	10–20	200–300
ML	Silts, sandy silts, gravelly silts, or diatomaceous soils	5–15	100–200
CL	Lean clays, sandy clays, or gravelly clays	5–15	100–200
OL	Organic silts or lean organic clays	4–8	100–200
MH	Micaceous clays or diatomaceous soils	4 8	100–200
CH	Fat clays	3–5	50–100
OH	Fat organic clays	3–5	50–100

emphasized that such a determination is only approximate in nature and that an exact measurement should be made for any particular soil. However, analyses made in advance of final designs may incorporate such approximate moduli.

The usual method for measuring k has been to place a plate on the ground, load the plate, and measure its penetration. From the initial tangent or secant slope of the stress-strain curve thus obtained a modulus is derived. Many different sizes of bearing area have been used in the past, and the results have been correlated by the U.S. Army Corps of Engineers. The Corps has compared the results of tests made with different sizes of plate and has presented them in the form of a curve (Fig. 11–10). Thus, the use of any one size of bearing plate in a particular instance can be identified with a standard result.

It may be noted that a precise determination of k is not, in practice, necessary, since a variation in k according to Eqs. (11–2) to (11–4) gives a smaller relative change in the slab's tensile stress. This means that the stress in the pavement is relatively insensitive to the foundation material.

Thus, in the design of a concrete pavement, the properties of the natural subgrade must be assessed by testing, when the construction thickness

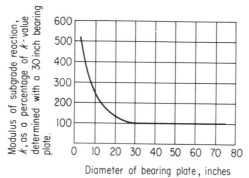

FIG. 11–10   *Relationship between modulus of subgrade reaction and diameter of bearing plate.* (Courtesy U.S. Corps of Engineers)

may be estimated, knowing the equivalent single-wheel load. The tensile stress in an estimated thickness of concrete slab can then be computed, knowing the load and the base course properties. The thickness of the concrete and the amount of reinforcing are then adjusted to ensure that these stresses lie within the tolerable limits of tensile stresses for concrete. A convenient slab size is chosen; in the case of highways, it determines the amount of mesh reinforcement to use. Knowledge of the wheel loads enables calculation of the size and spacing of dowel rods if they are to be used in the construction.

It is assumed in all design methods that the fill immediately underneath a pavement is adequately drained to at least a depth of 4 ft below the pavement surface in order that the good bearing properties of the fill can be sustained.

Since the loads sustained by an airfield or highway pavement are repetitive in nature, the possibility of soil failures or displacements occurring by a manner or mechanism analogous to that of fatigue failure in metals has been considered, although no design methods have been evolved to date. The problem is obviously a highly complex one, since the behavior of different types of soil under repetitious loading varies enormously. Research is being carried on to ascertain whether reasonable approximations to the behavior patterns of soil under repeated loadings can be obtained and unified into groups. At the moment, however, one must rely principally on empirical data involving little or no knowledge of the soil strength and considering only the satisfactory or otherwise behavior of test surfaces. For this reason the Western Association of State Highway Officials (WASHO) and more lately the American Association of State Highway Officials (AASHO) have built road test areas for the purpose of deriving more information on the behavior of different types of surfacing and subgrade of varying thickness under a wide variety of loading conditions.

PROBLEMS

11-1 A laboratory compaction test on a soil gives the following results:

DRY UNIT WEIGHT, pcf	MOISTURE CONTENT, % dry wt.
90	6.5
94	12.0
97	16.0
94	18.7
91	18.7
88	21.0

What is the specific gravity?

Answer Plot the compaction curve as in Fig. 11-1. The last three points indicate complete saturation and should be representative of a "zero air voids" condition. A check on any of these points shows specific gravity = 2.0.

11-2 Minimum compaction on a grading project is specified as 90 percent of the maximum dry unit weight (130 pcf) obtained in a standard laboratory test. Laboratory compaction was performed on material passing a $\frac{3}{4}$-in. sieve. A field inspector makes a sand-displacement density test. He finds 10 percent by dry weight of rocks larger than $\frac{3}{4}$-in. in his test hole. He removes these rocks from the material excavated and places them back in the hole before measuring its volume. The test yields a dry unit weight of 110 pcf, and excavation and recompaction are ordered. The contractor claims that exclusion of the rocks gave incorrectly low results. Prove that this claim is false. Specific gravity of the rock is 2.56.

Answer The contractor was right in that the inclusion of the rocks would have given a higher dry unit weight γ_d, which can be determined from $0.1\gamma_d/2.56 \times 62.4 + 0.9\gamma_d = 1$ ft^3, or $\gamma_d = 114$ pcf. However, in order to achieve the specified 90 percent compaction, the dry unit weight of the fine fraction should have been 117 pcf and that of the fine fraction plus 10 percent rock, using the same formula as above, should have been 120 pcf. Consequently, recompaction was necessary.

11-3 The rock content in a fill is 50 percent by dry weight. The minimum void ratio to which the rock can be compacted is 0.6. The maximum dry unit weight to which the soil fraction can be compacted is 120 pcf. What is the maximum dry unit weight to which the fill can be compacted? Specific gravity of the rock is 2.56.

Answer The maximum dry unit weight of rocks alone is

$$1.0/(1.00 + 0.6) \times 2.56 \times 62.4 = 100 \text{ pcf}$$

Maximum dry unit weight if remaining voids are filled with soil compacted to 120 pcf is

$$100 + 0.6/(1.0 + 0.6) \times 120 = 145 \text{ pcf}$$

However, this would require a rock content of $100/145 = 69$ percent. Therefore this maximum cannot be achieved and the maximum dry unit weight of the fill can be determined in the same way as in the preceding problem. For a 50:50 mixture the maximum dry unit weight is 137 pcf.

11-4 What would the maximum dry unit weight in Prob. 11-3 be if the rock content was 80 percent by weight?   *Answer*   145 pcf

11-5 The CBR test curve of Fig. 11-8 represents a test on a well-graded sandy gravel. If the minimum requirements for heavy commercial traffic (wheel loads up to 16,000 lb) call for 4 in. of asphalt concrete, how much base course material is required according to (a) Bureau of Public Works Group Index Method, and (b) Corps of Engineers Highway Design Curves?

Answer   (a) Group Index Method: $a = b = c = d = G = 0$; see Fig. 11-7: total thickness is 12 in., or 4 in. A.C. and 8 in. base course. (b) CBR is 67 percent for 0.1 in. penetration, 71 percent for 0.2 in. Use 67 percent; see Fig. 11-9. Total thickness required: 4.7 in., i.e., only minimal base course required. Note the savings gained by actually testing the soil!

REFERENCES

1. H. B. SEED and C. K. CHAN, Structure and Strength Characteristics of Compacted Clays, *J. Soil Mech. Found. Div. Am. Soc. Civil Engrs.*, vol. 85, SM5, 1959.

2. E. J. FELT, Factors Influencing Some of the Physical Properties of Soil-Cement Mixtures, *Proc. Highway Res. Board Bull.* 108, 1955.

3. T. W. LAMBE, A. S. MICHAELS, and Z. C. MOH, "Improvement of the Strength of Soil-Cement with Additives," Highway Research Board, 1957.

4. H. Q. GOLDER, Relationship of Runway Thickness and Under-carriage Design to the Properties of the Subgrade Soil, *Inst. Civil Engrs., Airport Paper* 4, London, 1946.

5. H. M. WESTERGAARD, New Formulas for Stresses in Concrete Pavements of Airfields, *Trans. ASCE*, 1948.

Index